AN

ELEMENTARY COURSE

IN

DESCRIPTIVE GEOMETRY

BY

SOLOMON WOOLF, A.M.,

Professor of Descriptive Geometry and Drawing in the College of the City of New York.

THIRD EDITION.

TABLE OF CONTENTS.

CHAPTER I.

GENERAL CONSIDERATIONS.

CHAPTER II.

THE LINE.

CHAPTER III.

LINES, PLANE FIGURES AND PLANES.

CHAPTER IV.

I. PARALLEL AND INTERSECTING PLANES.

CHAPTER IX.

ANGLES.

I. ANGLES BETWEEN RIGHT LINES.

II. ANGLES BETWEEN LINES AND PLANES.

III. ANGLES BETWEEN PLANES.

CHAPTER X.

CHANGE OF POSITION BY COMBINED MOTIONS.

CHAPTER XI.

SECTIONS.

I. ELEMENTARY SECTIONS.

II. OBLIQUE SECTIONS.

CHAPTER XII.

INTERSECTIONS.

I. ELEMENTARY INTERSECTIONS.

CHAPTER XIII.

TANGENTS AND NORMALS.

I. GENERAL CONSIDERATIONS.

II. TANGENTS AND NORMALS TO SURFACES.

III. TANGENTS TO RULED SURFACES.

ELEMENTS OF DESCRIPTIVE GEOMETRY.

CHAPTER I.

GENERAL CONSIDERATIONS.

1. In a limited sense, Descriptive Geometry may be defined to be a conventional method of representing on a plane objects which have three dimensions, so as to admit of an accurate determination of their size, form and position.

2. But position being relative, objects are in respect of this undeterminable unless referred to other objects, which, for constructive purposes, ought to be of the simplest character and in positions readily conceived and comprehended.

In Descriptive Geometry these latter objects are two or more planes,* termed the *planes of projection*.

3. These two planes intersect at right angles (Fig. 1), one being in the position of an ordinary wall and the other in the position of the

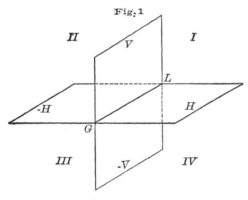

Fig. 1

floor. The upright plane, *V*, is the *vertical plane of projection*, and the other plane, *H*, at right angles to it, the *horizontal plane of projection*. The line *GL* in which they intersect is termed the *ground-line*.

4. The planes thus arranged are divided by their line of intersection, *GL*, into two parts each, and, in turn, divide space into four equal dihedral† angles. For convenience of reference the upper and lower

---

* Generally two ; three or more for special constructions.

† Δις, *two*, εδρα, *base* or *face*—angles bounded by two faces.

division of the vertical plane are lettered V and $- V$ respectively, the front and back divisions of the horizontal plane H and $- H$. With a like view to convenience the angles are numbered:

I, the angle *above H* and *in front* of V;
II, the angle *above H* and *behind V*;
III, the angle *below H* and *behind V*;
IV, the angle *below H* and *in front* of V.

THE POINT IN SPACE.

5. The planes of projections being thus given, the *projections*, so called, of any point in space are determined by letting fall from that point *perpendiculars to the two planes*.

Thus from a (Fig. 2) let fall the perpendicular aa''—termed the pro-

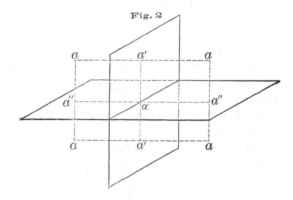

Fig. 2

jecting line—to the horizontal plane, H; the foot a'' of this line is the *horizontal projection* of the point.

The combined horizontal projections of the different points of any object constitute the horizontal projection of that object.

In like manner, a perpendicular, aa'—the projecting line—drawn from the point a to the vertical plane, V, marks with its foot, a', *the vertical projection* of that point.

The combined vertical projections of the different points of any object constitute the vertical projection of that object.

It is evident that the points a' and a'' thus determined are the projections of all points lying in the two projecting lines aa' and aa''.

6. The projecting lines aa', aa'' are respectively perpendicular to V and H; hence any plane which passes through them must likewise be perpendicular to those planes, and, by Geometry, to their line of intersection or ground-line, GL, which it cuts in the point α. The angle $a'\alpha a''$, which measures the dihedral angle of the two planes of projection, is thus a *right angle*.

7. The angles $aa'\alpha$ and $aa''\alpha$ being by construction right angles, the figure $aa'\alpha a''$ is a rectangle; whence it follows that—

(1) The distance *aa''* of the point *a* from *H* is equal to *a'α*, or the distance of the vertical projection *a'* *above* the ground-line, for the first and second angles, and *below* for the third and fourth.

(2) The distance *aa'* of the point *a* from *V* is equal to *a''α*, or the distance of the horizontal projection *a''* *in front* of the ground-line for the first and fourth angles, and *behind* for the second and third.

(3) The perpendiculars to *GL*, drawn through the projections *a'* and *a''*, intersect it in a common point, *α*.

The lines *a'α* and *a''α* are termed the *corresponding ordinates* of the point *a*, and *α* is its *ground-point*.

8. With this orthogonal* system of projection it is found that—

(1) All distances from the vertical plane are measured in their *true lengths* upon the horizontal plane.

(2) All distances from the horizontal plane are measured in their *true lengths* upon the vertical plane.

(3) By means of the combined projections the position of the point in space may be accurately determined.

9. Heretofore, the planes of projection, together with the constructions, have been considered solely with reference to their relative positions *in space*. But, in order to represent both projections of an object on one plane, the *drawing* plane—in other words, in order to render the

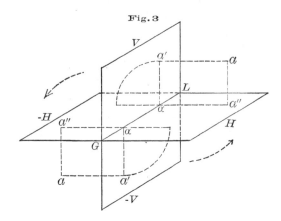

Fig. 3

drawings, thus made upon the planes at right angles to each other, available for practical purposes—some modification must be made in the position of these planes.

10. This is effected by revolving either plane of projection around the ground-line, *GL*, until it coincides with the other plane. Thus the plane *V* (Fig. 3) is ·rotated *backwards* in the direction of the arrows, until it coincides with the plane *H*, when the upper portion, *V*, rests upon the back portion, — *H*, of the horizontal plane, and the lower

---

* ὀρθός, *straight* or *right*, γωνία, *angle*—referring to the rectangular position of the projecting lines.

portion, — V, is covered by the front portion, H, of the horizontal plane.*

11. As the result of this rotation all that portion of the drawing plane which lies above the ground-line represents the negative horizontal, — H, and the superimposed vertical plane, V; while all below the ground-line is the negative vertical, — V, and the superimposed horizontal plane, H.

12. When the vertical plane is rotated backwards into the horizontal plane, the vertical projection a', and its ordinate, $a'\alpha$, are carried around with it, and maintain unaltered their position to GL. Hence, the ordinates $a'\alpha$ and $a''\alpha$, being now in the same plane, passing through the same point, α, and remaining perpendicular to the ground-line, are the *prolongations of each other*. Whence it follows that the two corresponding *projections of any point in space will always lie in the same line perpendicular to the ground-line*. Two points, one in the vertical and the other in the horizontal plane, cannot, therefore, be assumed at will to represent the projections of a point in space unless they fall in the same perpendicular to GL.

13. Conversely, the two projections (a', a'') *fix the position of the point in space*. For the line GL, being perpendicular to the two ordinates $a'\alpha$, $a''\alpha$, will likewise be so to the plane passing through them; hence, the planes V and H are by Geometry perpendicular to this same plane; or, reciprocally, the plane $aa'\alpha a''$ is perpendicular to V and H. If, then, the planes of projection be restored to their original position, and through the projections a' and a'' perpendiculars be erected to them, these perpendiculars will lie in the plane of $aa'\alpha a''$, and, of necessity, intersect in a point of which the projections will be (a', a'').

14. When, from the two projections, it is thus sought to determine the nature and position of the object in space, the planes of projection must always be imagined as restored to their rectangular position; it is only for the purpose of making the drawings or projections from the given data that the planes are made to coincide.

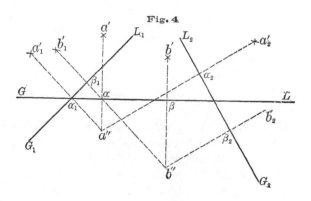

Fig. 4

---

*Long experience has shown the utility of requiring students to analyze the earlier problems by means of cards cut to represent the planes of projection in space.

CHANGE OF GROUND-LINE.

15. It is frequently desirable, nay, even necessary in practice, to employ more than one vertical plane. Thus, GL, G_1L_1 and G_2L_2 (Fig. 4) may represent the ground-lines of three distinct planes, all of which are perpendicular to the horizontal plane while assuming any position to one another.

If, then, a'' and b'' are the horizontal projections of two points in space, it is manifest that, by the application of Art. (12), a' and b', a_1' and b_1', a_2' and b_2' are the vertical projections of these same points on the three separate planes; $a'\alpha$, $a_1'\alpha_1$, $a_2'\alpha_2$ and $b'\beta$, $b_1'\beta_1$, $b_2'\beta_2$, measuring in each case the distances of the points above the common horizontal plane.

16. *Notation of the Point.*—Throughout this work the point in space will be designated by a small letter, and its projections by the same letter accented. Thus the point a in space is one whose projections are a' and a'' for the vertical and horizontal planes respectively.

On the plus planes the projections will be indicated by a cross, and on the minus planes by a small circle. Where the two projections coincide, which will take place in the second and fourth angles when the point lies in the bisecting plane, the cross will be enclosed within the circle.

POSITIONS OF THE POINT IN SPACE.

17. A point, in revolving around the ground-line, may assume the following general positions:

(1) On H, in front of V (Fig. 5).

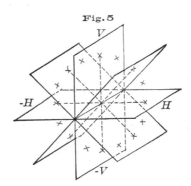

Fig. 5

(2) In the first dihedral angle, *below the bisecting plane,* in which case it is *nearer H* than V.

(3) In the bisecting plane, in which case it is *equidistant* from V and H.

(4) In the first angle, *above* the bisecting plane, in which case it is *nearer V* than H.

(5) In V above H.

(6) In the second dihedral angle, *above* the bisecting plane, in which case it is *nearer V* than $-H$.

(7) In the bisecting plane, equidistant from both.

(8) In the second dihedral angle, *below* the bisecting plane, when it is *nearer* — *H* than *V.*

(9) In — *H* behind *V.*

(10) In the third dihedral angle, *above* the bisecting plane, when it is *nearer* — *H* than *V.*

(11) In the bisecting plane, equidistant from both.

(12) In the third dihedral angle, *below* the bisecting plane, when it is *nearer* — *V* than — *H.*

(13) In — *V* below *H.*

(14) In the fourth dihedral angle, *below* the bisecting plane, when it is *nearer* — *V* than *H.*

(15) In the bisecting plane, equidistant from both.

(16) In the fourth dihedral angle, *above* the bisecting plane, when it is *nearer H* than — *V.*

(17) In the ground-line, when it is a point of both planes.

18. PROBLEM.—*To find the projections of a point, its distances from the planes of projection being given.*

Let the point (Fig. 6) be assumed to be half an inch from *H* and one inch from *V.*

V having been revolved (Art. 10), draw an indefinite horizontal line,

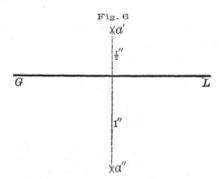

Fig. 6

the ground-line *GL.* Since the two projections must always lie in the same perpendicular to *GL* (Art. 12), draw an indefinite ordinate in that position, and lay off the distances half an inch above and one inch below *GL*; the points (*a′, a″*) thus determined are the projections sought. (Art. 8.)

19. *Analyze, from the given projections, the positions of the points in space* (Fig. 7).

20. PROBLEM.—*From the data* (Art. 18) *to find the projections of a point in the remaining dihedral angles.*

21. PROBLEM.—*To find the projections of any point lying in the bisecting planes of the first and fourth dihedral angles.*

22. PROBLEM.—*To find the projections of a point lying in either plane of projection.*

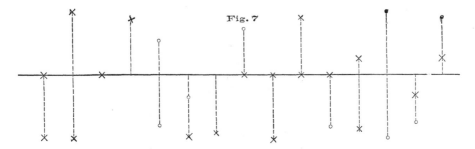

Fig. 7

23. PROBLEM.—*The primitive projections of a point being given, to find its new vertical projection upon any new vertical plane.*

(1) By Art. (12) the horizontal projection and the new vertical projection ought to lie in the same perpendicular to *GL*.

(2) The distance of the point from *H* remains unaltered. Hence if (a', a'') (Fig. 8) are the primitive projections of the point a, let fall upon the new

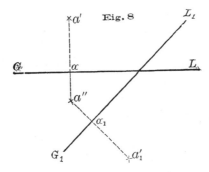

Fig. 8

ground-line G_1L_1 the ordinate $a''\alpha_1$, and lay off a distance α_1a_1' equal to $\alpha a'$, above or below G_1L_1 as the point a' may be above or below GL. The point a_1' is the new vertical projection sought.

CHAPTER II.

THE LINE.

24. A right line in space may have one of three positions to the planes of projection (Fig. 9):

(1) Parallel to *both*, *i.e.*, their line of intersection, or ground-line (*a*).

(2) Parallel to *one*, $\begin{cases} \text{perpendicular to the other } (b)\,; \\ \text{inclining to the other } (c). \end{cases}$

(3) Parallel to *neither*, $\begin{cases} \text{inclining at any angle } (d)\,; \\ \text{in a plane perpendicular to } GL\ (e)\,; \\ \text{intersecting } GL\ (f). \end{cases}$

25. As a line is a succession of points, its projection on any plane will be determined by projecting each point of the line on that plane. Thus, if from the different points of the line *AB* (Fig. 9, *a*) perpendiculars or projecting lines be drawn to *H*, their feet will indicate the horizontal pro-

Fig. 9

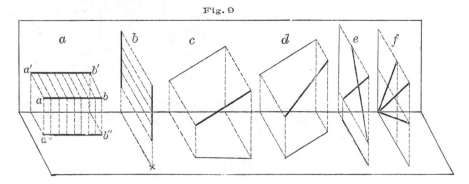

jections of those points, and the line *a″b″*, which passes through them, the horizontal projection of the line itself.

In like manner the vertical projection *a′b′* will be found by drawing projecting lines to the vertical plane.

26. This regular succession of parallel projecting lines will form a surface, which in the case of the right line is a *projecting plane*, but with the curved line a *projecting cylinder*.

In either case the projection of the line lies *in the line of intersection between the projecting surfaces and the planes of projection*.

27. Hence the projection of a right line is a right line,—with the single exception (Fig. 9, *b*),—and is obtained by passing through the line in space planes respectively perpendicular to the coördinate planes of projection.

28. As two points not consecutive geometrically determine the position of any right line, so the projections of any two of its points determine its projections.

29. In general, a right line will be fully determined by its two projections; for if the coördinate planes be restored to their rectangular position, and through each projection of the line a plane be erected perpendicular to the coördinate planes respectively, the two planes thus constructed will each contain the line in space which must be, consequently, their line of intersection. When both projections are, however, perpendicular to *GL* (Fig. 9, *e, f*), the projecting planes are also perpendicular to *GL*, and coincide; hence they will give no line of intersection, and fail to fix the line in space.

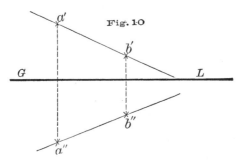

Fig. 10

30. PROBLEM.—*The projections of two points being given, to find the projections of the line which they determine.*

1st Case.—When the projections do not lie in the same perpendicular to *GL* (Fig. 10).

Let (*a', a''*) and (*b', b''*) be the given points. The projections must pass through the corresponding projections of the points (Art. 28), hence *a'b'* and *a''b''*, are the vertical and the horizontal projections sought.

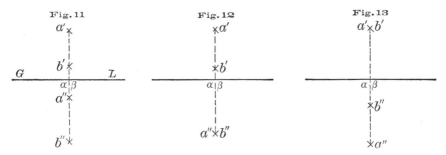

2d Case.—Where the projections of the points lie in a common perpendicular to *GL*.

If the projections (*a', a''*) and (*b', b''*) are separate points (Fig. 11), *a'b'* and *a''b''* are again the required projections of the line, but the projecting planes coincide and are perpendicular to *GL*.

If *a''* and *b''* coincide (Fig. 12), and *a'* and *b'* are separate, the entire line is projected horizontally *in a point*, and vertically in a line $a'b'$ perpendicular to *GL*.

If a' and b' coincide (Fig. 13), and a'' and b'' are separate, then the entire line is projected vertically *in a point*, and horizontally in a line perpendicular to *GL*.

The corresponding projections a', b' and a'', b'' cannot both respectively coincide, since the line in space cannot assume, at one and the same time, a vertical and a horizontal position to the coördinate planes.

31. Problem.—*Given the projections of two points, to find the projections of a third point of the line which they determine.*

1st Case.—Both projections of the line may incline to *GL*. Let (a', a'') and (b', b'') be the given points (Fig. 14). Draw the projections of the line passing through them. Any point c of the line ab in space gives a horizontal projection somewhere on the line $a''b''$, and a vertical projection somewhere on the line $a'b'$, since the projections of the line contain

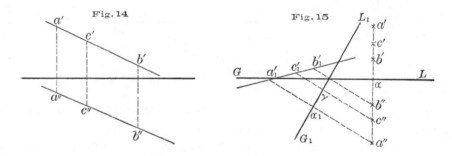

all the projections of its successive points. But as the projections of a point always lie in the same perpendicular to *GL*, the projections of the third point will be found by erecting a perpendicular to *GL* and marking its points of intersection (c', c'') with $(a'b', a''b'')$.

2d Case.—The line ab may lie in a plane perpendicular to *GL*.

As in the preceding case the point c will give projections which are points of the projections of the line ab. But as the two projections lie in a common perpendicular to *GL*, the ordinate drawn from any point coincides with that perpendicular, and hence determines no point of intersection. (Fig. 15.)

Assume, then, a new ground-line G_1L_1 which shall not be parallel to the primitive one, and find the new vertical projections (a_1', b_1') and the projection of any intermediate point c_1' of the line joining them. Set off the distance γ_1c_1' thus obtained from α to c' and (c', c'') are the required projections.

In practice, whenever the conditions of the problem render it practicable, the new ground-line ought to be taken at right angles to the primitive one (Fig. 16), thereby avoiding more or less complex constructions.

3d Case.—The line may be perpendicular to either plane.

In either case the solution is immediate. Should the line be vertical (Fig. 17), c'' will coincide with the horizontal projection $a''b''$, wherever the projection c' may be assumed.

Should the line be horizontal (Fig. 18), c' will coincide with the vertical projection $a'b'$, wherever c'' may be assumed.

32. The preceding case leads to the following

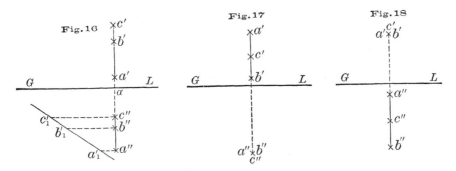

PROBLEM.—*A line being given, to find upon that line* (1) *the vertical projection of any given point;* (2) *the horizontal projection of any given point.*

(1) The distance of any point from the horizontal plane is the measure of the distance of the vertical projection from GL; hence, to determine upon the vertical projection of the line that point whose distance is known, draw a line $c'd'$ (Fig. 19) parallel to GL and at the required distance from it; the point of intersection (a', a'') is the one sought.

(2) A similar construction, $c''d''$, determines by its point of intersection (a', a'') the point at a given distance from V.

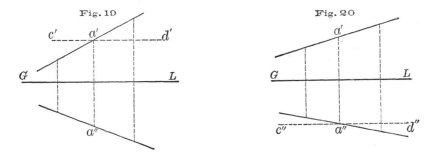

33. PROBLEM.—*A line being given, to find upon that line* (1) *the point whose distance from H is a minimum;* (2) *the point whose distance from V is a minimum.*

(1) From the preceding case it is evident that the nearer the extremity of the projection $a'b'$ approaches GL, the nearer it is to H; hence, when the projection is prolonged until it intersects GL, the point of intersection h' is the vertical projection of the required point (Fig. 21) from which the horizontal projection h'' may be determined by means of the ordinate.

The point (h', h'') whose distance from H is a minimum is that in which the line *ab* in space pierces *H*, and is termed the *horizontal trace* or *piercing-point* of that line.

(2) The nearer the extremity of the projection *a''b''* (Figs. 21, 22, 23) approaches *GL* the nearer is the line in space to *V*; hence, when the projection is prolonged until it intersects *GL*, the point of intersection *v''*

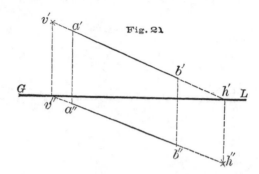

Fig. 21

is the horizontal projection of the required point from which the vertical projection *v'* may be determined by means of the ordinate.

The point (*v'*, *v''*) whose distance from *V* is a minimum is that in which the line *ab* in space pierces *V*, and is termed the *vertical trace* or *piercing-point* of that line.

When the given line lies in a plane perpendicular to *GL* (Figs. 22, 23),

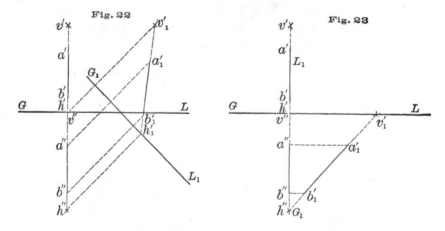

Fig. 22   Fig. 23

determine the piercing-points on a new vertical plane, and transfer the measurements thus obtained upon the primitive planes of projection.

In Fig. 23 the new vertical plane coincides with the plane of the line.

With geometrical lines capable of unlimited extension these two piercing-points are peculiarly fitted to fix the position of the line in space to the planes of projection.

34. The preceding case may assume the following form:

PROBLEM.—*To find the traces or piercing-points of a line the projections of which are given.*

35. Applying the principle of piercing-points to the cases of Art. (24), it will be seen that—

When the line in space intersects *neither* plane of projection it must be parallel to both, and hence to their line of intersection, or *GL*. In such a position *both projections are parallel to GL*, since under any other supposition they would intersect it and thus make the line in space pierce the coördinate planes, which is contrary to the conditions of the problem.

Nine cases are to be noted (Fig. 24); *four* as the line lies in either dihedral angle, *four* as it lies in either coördinate plane, and *one* as it lies in the ground-line. When the line is equidistant from the coördinate planes it lies in the bisecting planes; hence in the second and fourth angles the projections coincide.

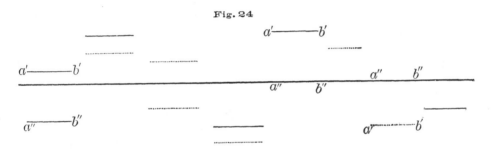

Fig. 24

36. [*Graphical Characteristics of Lines.*—As the right line is fixed in position by any two points which are not consecutive, in a similar way its projections will be indicated. Thus *ab* will express a line in space, and *a'b'*, *a"b"* its vertical and horizontal projections respectively.

It is customary to distinguish between the various lines which are the products of the constructions, as, for instance, between those which are seen or concealed, and, again, between those which indicate the given data and the results, and those which are simply the means of their attainment. Accordingly they may be classified and delineated as—

(1) *Principal lines*, the graphical representations of the data and the results attained therefrom, drawn *full* when seen and *dotted* when hidden, whether by the planes of projection, portions of an object of which they form a part, or by the interposition of other objects.

(2) *Construction lines*, employed as auxiliaries in determining the required solution or connecting corresponding projections of the several points, lines, etc., and drawn as broken lines composed of short dashes.]

37. When the line in space intersects but *one* of the coördinate planes it is, of necessity, parallel to the other, whatever the position it assumes to the plane it intersects. Should it be a vertical line, its horizontal projection is reduced to a *point* (Fig. 9, *b*), since the projecting lines and the line in space coincide. The vertical projection is a line *perpendicular to the ground-line*, since the vertical projecting plane and the vertical plane of projection are both perpendicular to *H*.

There are *three* general positions to be noted for this case, according as the line is situated in front of *V*, in that plane or behind it. (Fig. 25.)

Three similar positions (Fig. 26) are indicated when the line is perpendicular to *V*, according as it is situated above, in or below *H*.

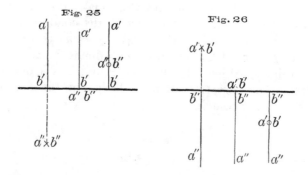

38. *A line in space parallel to either plane of projection and inclining to the other is projected on the former in a line parallel and equal to itself, and on the latter in a line parallel to GL.*

Should the line be parallel to *V* (Fig. 27, 1), then the vertical projection *a'b'* will be parallel to the line *ab* in space, since these two lines, lying in the same projecting plane, cannot, by the conditions of the case, intersect. They are equal inasmuch as they are the opposite sides of the rectangle *aba'b'*.

The horizontal projection *a''b''* will be parallel to *GL* inasmuch as the

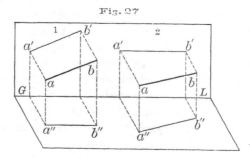

projecting plane and *V* are parallel, and hence are cut by *H* in parallel lines.

Should the line *ab* be parallel to *H* (Fig. 27, 2), by similar reasoning it may be shown that the horizontal projection *a''b''* will be parallel and equal to *ab*, and the vertical projection *a'b'* parallel to *GL*.

39. Fig. 28 exhibits the various general positions which this case may assume as the line *ab* in space is located above, in or below *H*, in front of, in or behind *V*.

(1) Above *H*, oblique to *V*;   (4) In front of *V*, oblique to *H*;
(2) In *H*, oblique to *V*;   (5) In *V*, oblique to *H*;
(3) Below *H*, oblique to *V*;   (6) Behind *V*, oblique to *H*.

40. Any line lying in the bisecting planes of the dihedral angles will be projected in the first and third angles in lines which incline in the same direction and at the same angle to *GL*, and in the second and fourth in lines which coincide.

From this it results that any line parallel to a bisecting plane—that is, parallel to any line of that plane—will be projected in the first and third angles as above, and in the second and fourth in parallel lines.

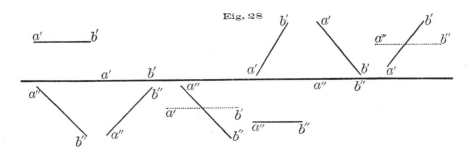

Fig. 28

41. When the line in space intersects *both* coördinate planes it inclines to both, *and gives projections which are shorter than itself and which incline to GL.*

Thus, in Fig. 29 the line *ab* is the hypothenuse of the right-angled triangles *abc* and *abd*, hence is greater than the bases *ac* and *bd* or their equivalents, the projections $a''b''$ and $a'b'$.

Again, the projecting planes of the line *ab* incline to *GL*, cut the coördinate planes in inclining lines, and hence give inclining projections.

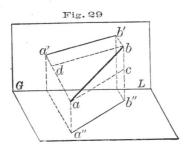

Fig. 29

Exceptions arise when the line lies in a *profile* plane perpendicular to *GL* (Fig. 9, *e*, *f*), in which cases both projections lie in the traces, and consequently are perpendicular to *GL*.

In case *e* there are four positions for the line, as a portion of it is intercepted in any one of the four dihedral angles.

In case *f*, should the line lie in the bisecting plane, its projections in the second and the fourth angles will respectively coincide.

42. Analyze the positions of the lines given by their projections (Fig. 30).

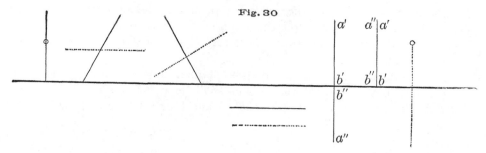

Fig. 30

43. PROBLEM.—*To find the projections of a right line parallel to GL, $1\frac{1}{2}''$ long, $\frac{3}{4}''$ from H and $\frac{1}{2}''$ from V.*

(1) Draw lightly an indefinite ordinate perpendicular to *GL* (Fig. 31).

(2) Lay off upon this line a distance of three quarters of an inch above and half an inch below *GL*, thus determining the projections (a', a'').

(3) Draw two full lines, $a'b'$, $a''b''$, $1\frac{1}{2}''$ in length, through these points and parallel to *GL*; the lines so drawn are the projections required.

From the given data determine the projections in the remaining three angles.

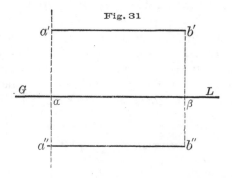

Fig. 31

44. PROBLEM.—*To find the projections of a right line $1''$ long, $\frac{1}{4}''$ from either plane and perpendicular to it.*

Let the line be vertical (Fig. 32, 1).

(1) Draw an indefinite ordinate perpendicular to *GL*.

(2) Mark upon this line two points, a', b', respectively indicating in their distances from *GL* the heights of the two extremities of the line above *H*.

(3) On this same ordinate lay off a distance $\alpha a''$ equal to the distance of the line from *V*. Then will ($a'b'$, $a''b''$) be the projections required.

From the given data determine the projections in the other dihedral angles.

45. PROBLEM.—*To find the projections of a line parallel to one plane of projection, and inclining to the other.*

Let the line be $1\frac{1}{4}''$ long, parallel to *V*, and inclining at any angle α to *H*; $1''$ from *V* and $\frac{1}{4}''$ from *H*.

(1) Draw an indefinite ordinate (Fig. 32, 2), and mark the projections *a'* one quarter of an inch above, and *a''* one inch below, *GL*, thus indicating the projections of the *lower* extremity of the line in space.

(2) Draw through *a'* the vertical projection *a'b'*, one and a half inches long and making the angle *α* with *GL*.

(3) Through *a''* draw the horizontal projection *a''b''* parallel to *GL*, and limited by the ordinate let fall from *b'*.

From the given data draw the projections in the other dihedral angles.

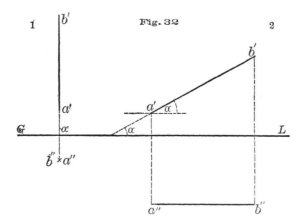

Fig. 32

46. PROBLEM.—*To find the projections of any line, its piercing-points being given.*

Let *a'* and *b''* be the given piercing-points (Fig. 33).

The point *a'* being in *V* is its own vertical projection, its horizontal projection, *a''*, falling in *GL*. Similarly, *b''* being a point in *H* is its own horizontal projection, its vertical projection, *b'*, falling in *GL*. Joining the

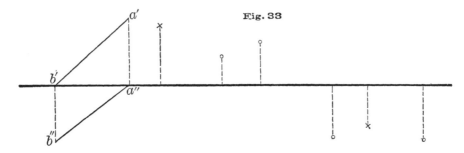

Fig. 33

corresponding projections *a'* and *b'*, *a''* and *b''*, (*a'b'*, *a''b''*) are the projections required.

Solve the remaining three cases (Fig. 33).

47. PROBLEM.—*A line being given by its projections, to find a point of that line whose distances from the coördinate planes shall be in a given ratio.*

Let the ratio be as *x* : *y*, and let (*a'b'*, *a''b''*) be the given line (Fig. 34). Draw any perpendicular to *GL* intersecting the projections in the points c', c'', and divide $c'c''$ in such a way that the two parts shall bear to each other the given ratio. It is manifest that this may be effected in two ways, as the point of division, α, lies between or beyond the points c', c''. The proportions then will be $c''\alpha : c'\alpha :: x : y$ for α intermediate, and $\alpha_1c'' : \alpha_1c' :: x : y$ for α_1. Join α and α_1 with d', the point in which the **pro-**

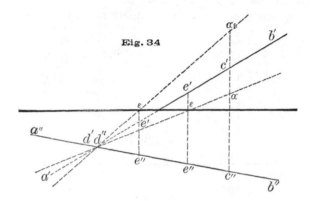

Fig. 34

jections cross, and draw ordinates through the points ϵ and ϵ where these lines cut GL; the points (e', e''), in which these ordinates intersect the given projections, are the projections of the required point.

48. PROBLEM.—*A line being given by its projections, to find a point of that line whose distances from the coördinate planes shall be equal.*

As in the preceding case, draw any perpendicular $c'c''$ to GL (Fig. 35).

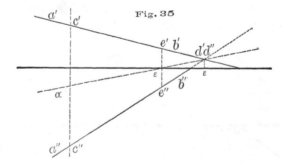

Fig. 35

Mark its intersections c', c'' with $a'b'$, $a''b''$, and divide $c'c''$ so that the parts shall be in the given ratio of $1 : 1$. Join the division-point, α, with d', and erect an ordinate at the point where the line thus obtained cuts GL; the points (e', e'') are the projections required.

In this case the second division-point is at infinity; hence the line joining it with d' is a perpendicular to GL, thus giving (d', d'').

CHAPTER III.

LINES, PLANE FIGURES AND PLANES.

49. Two lines in space may assume two general positions to each other: they may either lie in the same plane, or they may have such a position that no plane can be passed through them. In the first case they may either intersect or be parallel; in the second case they can affect neither of these positions.

PARALLEL LINES.

50. Two *parallel lines in space give projections that are parallel*, since their projecting planes are parallel, and hence cut the coördinate planes in parallel lines.

Should either pair of projections be reduced to points, the lines in space are parallel, since they are perpendicular to the same plane.

51. Conversely, *when the projections of the same name are parallel, the lines in space are parallel*, unless they lie in planes perpendicular to *GL*,

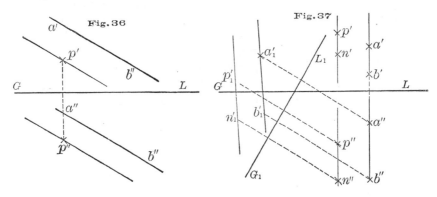

in which case the true position can be determined by a change of ground-line.

52. PROBLEM.—*Through a given point to pass a line parallel to a given line.*

Three cases are presented for solution:

(1) Both projections of the line may incline to *GL*.

Let (p', p'') be the given point (Fig. 36).

Through p' and p'' draw parallels respectively to $a'b'$, $a''b''$, the projections of the given line.

(2) The line may lie in a plane perpendicular to *GL* (Fig. 37).

Assume a new ground-line, G_1L_1, and determine the new vertical projections p_1' and $a_1'b_1'$ of the given point and line. Through p_1' draw $p_1'n_1'$ parallel to $a_1'b_1'$, and find its horizontal projection $p''n''$ parallel to $a''b''$. Through p' draw a parallel to $a'b'$, and determine the point n' by laying off a distance above GL equal to the distance of n_1' from G_1L_1.

(3) The line may be perpendicular to either coördinate plane, in which case the solution is at once obtained.

INTERSECTING LINES.

53. *Two lines in space which intersect give projections which intersect in points which lie in the same perpendicular to GL.*

For the point of intersection (Fig. 38) being common to both lines, its·

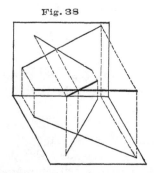

Fig. 38

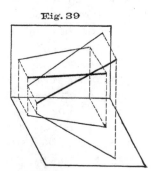

Fig. 39

projections must likewise be common to the two projections and follow the law which governs the projections of any point (Art. 12).

Conversely, when the like projections of the lines intersect in points lying in a common perpendicular to GL, the lines in space intersect.

Whence it follows that intersecting projections cannot be assumed at will to represent the projections of intersecting lines unless the points of intersection are in conformity with the above condition (Fig. 39).

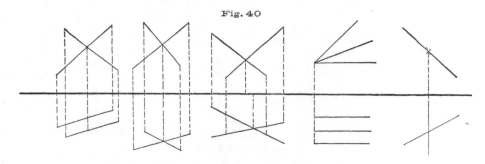

Fig. 40

54. PROBLEM.—*Analyze the positions of the lines in space whose projections are given* (Fig. 40).

55. Should either or both lines lie in a plane perpendicular to GL, a new ground-line will be required to determine their position to each other. Whence arises the following:

PROBLEM.—*To determine the position of two lines when either or both lie in a plane perpendicular to GL.*

Let $(a'b', a''b'')$ and $(c'd', c''d'')$ be the given lines (Fig. 41). Assume a new ground-line, G_1L_1, and find the new vertical projections. The lines intersect, inasmuch as p_1', the point of intersection on the new vertical plane, and p'' lie in a common perpendicular to *GL*.

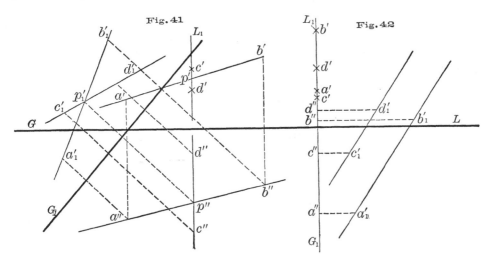

Fig. 42 exhibits the case in which the two lines *ab* and *cd*, both lying on a plane perpendicular to *GL*, are found, by means of the new vertical projections, to be parallel to each other.

56. PROBLEM.—*To determine the projections of any line connecting two given lines which intersect.*

Let $(a'b', a''b'')$ and $(c'd', c''d'')$ be the given lines (Fig. 43).

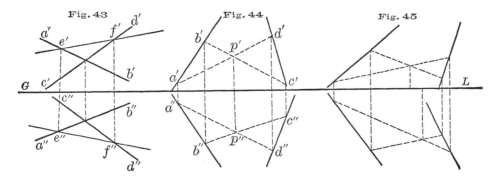

Assume any vertical projection $e'f'$ of the required line; the points of intersection e'' and f'' must lie in perpendiculars to *GL*, drawn through e' and f' respectively. The line $e''f''$ is the horizontal projection sought. This case admits of an infinite number of solutions.

57. PROBLEM.—*To determine the relative position of two lines whose projections do not intersect within the limits of the drawing.*

Let $(a'b', a''b'')$ and $(c'd', c''d'')$ be the two lines (Fig. 44).

Assume, as in the preceding case, any two lines $a'd'$ and $b'c'$ intersecting the given lines, and find their horizontal projections $a''d''$ and $b''c''$. Should the given lines intersect in space, the assumed lines will lie in their common plane and also intersect, and hence their points of intersection (p', p'') fall in the same perpendicular to *GL*.

Should the lines in space not intersect (Fig. 45), the points in which the projections cross each other will not lie in the same perpendicular to *GL*.

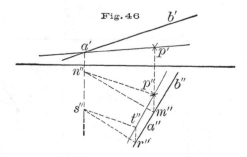

Fig. 46

58. PROBLEM.—*To draw the projections of a line passing through a given point and intersecting a given line, when either projection of the point of intersection lies beyond the limits of the drawing.*

Let $(a'b', a''b'')$ be the given line (Fig. 46), and (p', p'') the given point.

Assume $p'a'$, the vertical projection of the line required. Construct any triangle $p''m''n''$, one point lying in the horizontal projection $a''b''$, and the other in the ordinate passing through the point of intersection a'. Construct a second triangle in which the side $r''s''$ is parallel to $m''n''$, and through r'' and s'' draw parallels respectively to $m''p''$ and $n''p''$, thus determining the point t''. The line joining p'' and t'' is the horizontal projection sought.

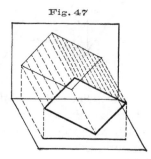

Fig. 47

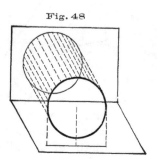

Fig. 48

PLANE FIGURES.

59. When the intersecting lines exceed two in number and lie in the same plane, they may enclose a *plane figure* whose projections can be readily determined by the preceding explanations.

In projecting any plane figure, whether rectilinear or curvilinear, the projecting planes form prismatic or cylindric surfaces (Figs. 47, 48) whose intersections with either plane of projection mark the projections.

60. PROBLEM.—*To find the projections of any plane figure parallel to the vertical plane.*

Assume the regular octagon to be the given figure (Fig. 49).

Conceive the projecting planes of the different sides; they form in space a prism the base of which is the given object. When such an

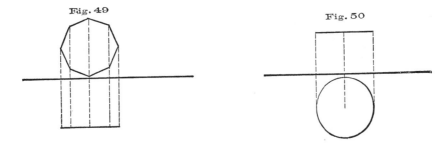

Fig. 49   Fig. 50

imaginary prism is cut by *V*, the section—that is, the vertical projection—is, by Geometry, equal to that base. The projecting plane to *H* and the plane of the object coincide and are parallel to *V*; hence the horizontal projection is a right line parallel to *GL*.

61. PROBLEM.—*To find the projections of a circle parallel to the horizontal plane.*

In this case the projecting surface becomes a cylinder (Fig. 50), the base of which is the circle in space. When this surface is intersected by *H*, the section—that is, the horizontal projection—is, by Geometry, a circle. The projecting plane to *V* and the plane of the circle coincide and are parallel to *H*; hence the vertical projection is a right line parallel to *GL*.

From the two preceding cases it necessarily follows that—

Any plane figure parallel to one of the coördinate planes gives a projection on that plane parallel and equal to itself, and on the other plane a right line parallel to GL.

62. PROBLEM.—*To determine the projections of a rectangle the plane of which is perpendicular to V and inclines to H at any given angle.*

Let *α* be the given angle, and let the longer side of the rectangle be perpendicular to *V* (Fig. 51), thus making the smaller side parallel to that plane.

Since the projecting surface to *V* and the plane of the object coincide, draw the vertical projection *a'b'* equal to the smaller side and making the given angle *α* with *GL*. Draw the ordinates, and lay off on them at the required distance from V the full length $a''a''$ of the longer side, that side being by the conditions of the problem parallel to H. Connect the extremities of these lines to complete the horizontal projection.

63. PROBLEM.—*To determine the projections of a circle the plane of which is perpendicular to H and inclines to V at any given angle.*

Since the projecting surface to H and the plane of the circle coincide, draw the horizontal projection $b''d''$ a right line equal to the diameter and making the given angle α with GL. Assume a new vertical plane parallel to the circle, and find the new vertical projection. Determine the horizontal projections b'', a'', d'', etc., of points of the circumference, and find the vertical projections by the usual method. The section of the projecting cylinder is, by Geometry, the curve of an ellipse.

64. Should the plane of the object be perpendicular to GL, both pro-

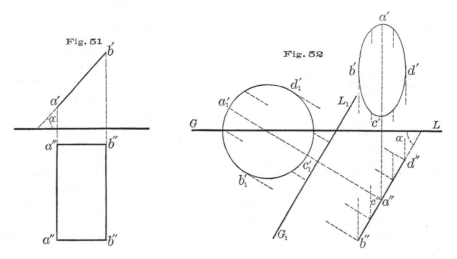

jections will be right lines perpendicular also to that line, since the projecting planes and the plane of the object coincide in space.

PLANES.

65. A plane is a surface which is generated by the motion of a right line in such a way that each of its points describes in space a line parallel to some second fixed right line. The moving line is termed the *generatrix*, the fixed one the *directrix*.

66. The position of a plane in space may be determined by two lines either parallel or intersecting, by a line and a point not in that line, or by three points not in the same right line.

67. As planes are of indefinite extension, they must intersect one or both planes of projection. The lines in which they so intersect are termed the *traces*, and in ordinary practice are the means whereby the planes are determined in position.

68. The plane in space may assume the following general positions to the coördinate planes: it may—

(1) Pass through the ground-line, giving no traces, and hence indeterminable except by the use of a new vertical plane (Fig. 53);

(2) Be parallel to one, giving but one trace parallel to *GL* on that plane to which it is not parallel (Fig. 54);

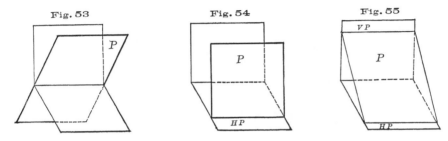

(3) Be parallel to *GL* and incline to both planes, when both traces will be parallel to that line (Fig. 55);

(4) Be perpendicular to *GL*, when both traces will be perpendicular to that line—the *profile plane* (Fig. 56);

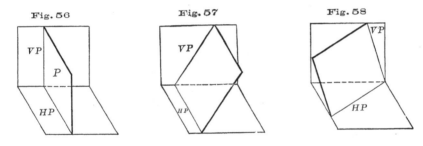

(5) Be perpendicular to one plane and incline to the other, when one trace will be perpendicular to *GL* and one inclined to it (Fig. 57);

(6) Incline to both planes, other than in the first and third cases, when both traces will incline to *GL* (Fig. 58).

The graphical representations of these cases are shown in Fig. 59.

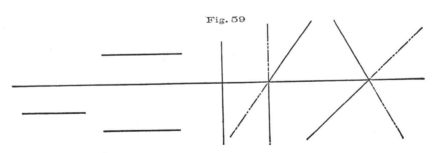

Fig. 59

69. The traces being lines in the coördinate planes are their own projections on the planes in which they are contained, their other projections being in *GL*.

70. Besides the traces the following are characteristic lines of the planes in space:

(1) The *horizontals*, or the lines parallel to *H* (Fig. 60).

(2) The *verticals*, or the lines parallel to *V* (Fig. 61).

(3) The *lines of greatest declivity*, which measure the angle of the plane in space with either plane of projection (Fig. 62).

71. With reference to these lines it is to be observed that—

(1) The horizontals are necessarily parallel to the horizontal trace; otherwise they would intersect it and, hence, the horizontal plane.

(2) The verticals are necessarily parallel to the vertical trace for a similar reason.

(3) The lines of greatest declivity are perpendicular to the respective traces (Fig. 62).

Thus, the line *po* drawn perpendicular to the horizontal trace *HN* measures the greatest declivity of the given plane with *H*. For, if from *p* we draw *pr*, inclining to the horizontal trace, and *pp″* perpendicular

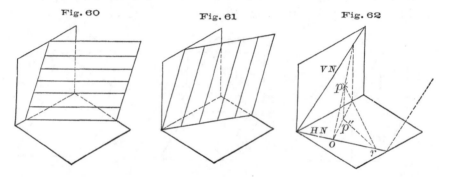

Fig. 60   Fig. 61   Fig. 62

to *H*; then *pr* > *po*, and *p″r* > *p″o*, while *pp″* remains unchanged. But in two right-angled triangles having the same perpendicular, *the greater the base, the smaller the angle at the base.*

72. The line *pp″* being perpendicular to *H*, the triangle *pp″o* lies in a plane perpendicular both to *H* and to the horizontal trace *HN*. Hence the angle of greatest declivity, *pop″*, will be determined by passing a plane at right angles to the trace on that plane to which the inclination is sought.

Whence it follows that *the traces do not necessarily measure the angles of inclination.*

73. The traces of a plane serve to indicate its position in all cases except that in which they coincide with *GL* (Fig. 53), since they represent the projections of two right lines of that plane which must either intersect or be parallel to each other.

74. When the traces of a plane are not parallel they must intersect each other in a point in *GL*, since the coördinate planes and the given plane form a solid angle whose vertex is the point of intersection or the piercing-point of *GL* on that plane.

75. *Notation of the Plane.*—Planes are usually represented by capital letters, and their traces by the prefixes *H* and *V*. Thus (Figs. 53–58), *P* denotes the plane in space, and *HP* and *VP* the horizontal and vertical traces respectively.

When a plane is given by two right lines it is indicated by the letters of the lines; thus, plane (*ab, cd*), which signifies the plane determined by the right lines *ab* and *cd*. In a similar way, the plane (*ab, c*) indicates a plane determined by the right line *ab* and point *c*. Lastly, the plane (*a, b, c*) expresses one whose position is fixed by the three points in question.

Visible traces are drawn full, concealed traces by a short dash and two dots alternately.

Auxiliary planes are represented by traces drawn with a short dash and dot alternately.

76. Problem.—*To analyze the position of a plane given by its traces* (Fig. 63).

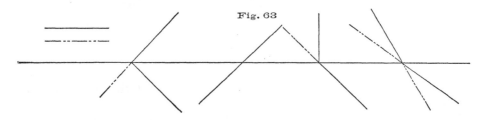

Fig. 63

77. Problem.—*To pass through a line in space a plane perpendicular to either coördinate plane.*

If perpendicular to *H* (Fig. 64), the required plane will coincide with the projecting plane of the line to *H*; hence its horziontal trace, *HN*,

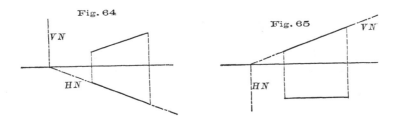

Fig. 64

Fig. 65

will contain the horizontal projection of the line, and its vertical trace, *VN*, will be perpendicular to *GL*.

If perpendicular to *V* (Fig. 65), then, for similar reasons, the vertical trace will contain the vertical projection of the line, and the horizontal trace will be perpendicular to *GL*.

78. Problem.—*Given three points not in the same right line, to find any line of the plane which they determine.*

Applying a principle of Geometry that "any line lying in a plane intersects all other lines of that plane except those to which it is parallel," the following cases may be distinguished:

(1) The plane (a, b, c) in which the projections of the three points do not coincide (Fig. 66). Join the points by three lines ($a'b'$, $a''b''$), ($b'c'$, $b''c''$) and ($c'a'$, $c''a''$), and take any intermediate points, as (o', o'') and (p', p''), on the lines so determined; the line connecting these points is the line required.

Again, if through any point as (o', o'') of the line ab (Fig. 67) a parallel

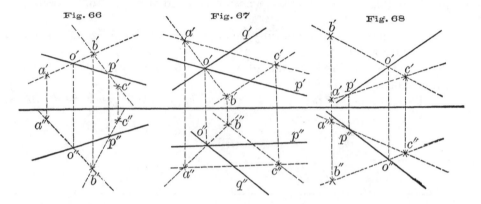

be drawn to either of the other two lines, a line of the plane will have been found.

(2) In Fig. 68 the two lines ($b'c'$, $b''c''$) and ($c'a'$, $c''a''$) alone admit of the immediate use of an intermediate point; the solution, however, may be effected as in the preceding cases.

Should the horizontal projections a'' and b'' coincide (Fig. 69), the

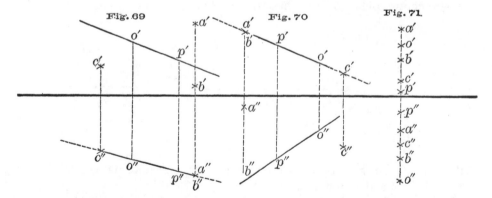

line joining them is *vertical* in position; hence the plane passing through the given points is likewise vertical, and every line lying in that plane is projected on H in its horizontal trace. From this it follows that any two points as (o'', p'') determine a line of the plane.

Should the two projections a' and b' coincide (Fig. 70), the line joining them is perpendicular to V; hence the plane passing through the given points is likewise perpendicular to V, and every line lying in that plane is projected on V in its vertical trace. From this it follows that any two points (o', p') determine the vertical projections of any line of that plane.

(3) Should the three points given lie in a profile plane, the plane being perpendicular to both coördinate planes, every line of that plane is projected upon the traces thereof (Fig. 71). Hence, any two points (o', o'') and (p', p'') assumed in these traces determine a line of the plane.

79. PROBLEM.—*Given the projections of three points not in the same right line, to find any point of the plane which they determine.*

Draw any line of the plane, as in preceding cases, and find any point of that line by the usual methods.

80. PROBLEM.—*Given one projection of a line lying in a given plane, to find the other projection.*

Let P be the given plane, and $h''v''$ the given projection (Figs. 72,

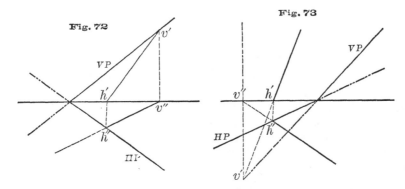

Fig. 72

Fig. 73

73). The piercing-points of the line must lie in the traces of the plane and be common to the projections of the line; hence, the horizontal piercing-point h'' falls at the intersections of HP and $h''v''$. The vertical piercing-point, being a point of the vertical plane and of the vertical trace VP, is projected horizontally at v''; the ordinate drawn from this

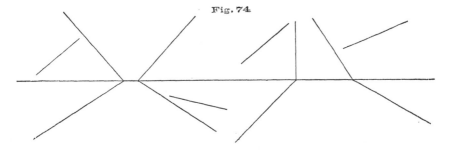

Fig. 74

point intersects VP in the point v', the vertical piercing-point of the given line. As any point in H is vertically projected in GL, $h'v'$ is the required vertical projection.

81. PROBLEM.—*Solve the cases of Fig. 74 under Art.* 80.

82. PROBLEM.—*Given the traces of a plane, to find the projections* (1) *of any horizontal,* (2) *of any vertical.*

(1) Let *VP* and *HP* be the given traces (Fig. 75).

Since the horizontal is a line parallel to the horizontal trace (Art. 71), and hence to *H*, its horizontal projection $v''h''$ is projected parallel to *HP*, and its vertical projection parallel to *GL*. Assume, then, any horizontal projection, determine its vertical piercing-point (v', v''), and draw $v'h'$ parallel to *GL*.

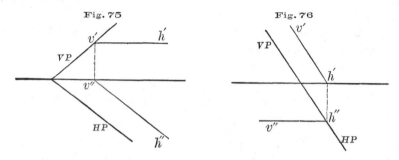

(2) In like manner to determine the vertical, draw any line $h'v'$ (Fig. 76) parallel to *VP*, find (h', h'') the horizontal piercing-point, and draw $h''v''$ parallel to *GL*.

83. PROBLEM.—*The plane being given by three points not in the same right line, to find a vertical or a horizontal.*

Draw the horizontal or vertical projection parallel to *GL* and proceed as in Art. 78.

84. PROBLEM.—*Given one trace of a plane and the projections of any point in that plane, to find the other trace.*

Let *VP* be the given trace (Fig. 77), and (a', a'') the given point. Through the point draw the vertical ($h'v'$, $h''v''$), and find its piercing-point h''; the required trace *HP* passes through it.

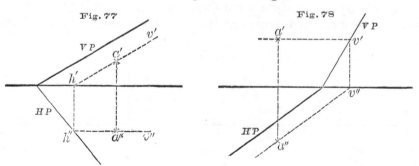

In Fig. 78 the vertical trace *VP* has been determined by means of the horizontal ($a''v''$, $a'v'$).

85. PROBLEM.—*To find the traces of a plane passing through two intersecting lines.*

Let $(a'b', a''b'')$ and $(c'd', c''d'')$ be the two lines intersecting in (o', o'') (Fig. 79). As the piercing-points of lines are always in the traces of the plane which contains them, find the piercing-points and pass the traces *VP* and *HP* through their vertical and horizontal projections respectively. When prolonged they intersect in *GL*, and thus verify the construction.

86. PROBLEM.—*To pass a plane through three given points not in the same right line.*

Let *a, b, c* be the given points in space (Fig. 80).

Join them two and two by lines, and find the piercing-points thereof; the traces of the required plane *VP* and *HP* pass through them.

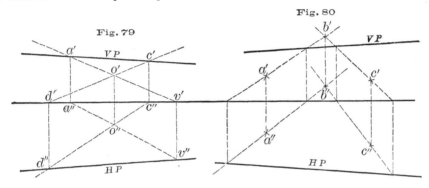

Fig. 79 Fig. 80

If the construction has been accurately determined, the three piercing-points of the same name will lie in the respective traces, and the traces themselves intersect in *GL*. A special case arises when the line joining two of the points is parallel to either or both planes of projection.

Should the line be parallel to *V* or *H*, then the vertical or the horizontal trace of the required plane will be parallel to the vertical or horizontal projection of that line.

Should the line be parallel to *GL*, the traces will also be parallel to that line.

87. PROBLEM.—*To pass a plane through a given point and line.*

A line passed through the given point parallel to or intersecting the given line lies in the required plane. The piercing-points of the auxiliary and given lines determine the traces of the plane.

CHANGE OF GROUND-LINE.

88. PROBLEM.—*Given a plane by its traces, to determine a new vertical trace by a change of ground-line.*

Let (VP, HP) be the given traces, and G_1L_1 the new ground-line (Fig. 81).

The new vertical plane intersects the traces of the given plane in two points, h'' and a point whose horizontal projection is v''. Lay off the distance $v''v_1'$ equal to $v''v'$, and $h''v_1'$ is the new vertical trace required.

Fig. 82 illustrates the case in which the new vertical plane has been taken at right angles to the primitive vertical, and Fig. 83 the case in

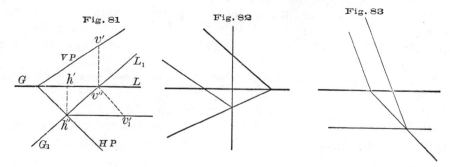

Fig. 81   Fig. 82   Fig. 83

which the new vertical plane has been taken parallel to the primitive vertical, the new vertical trace being for this position parallel to the given vertical trace.

CHAPTER IV.

I. PARALLEL AND INTERSECTING PLANES.

89. Two planes may assume two general positions to each other: they may be parallel or they may intersect.

With the parallel system, whatever the position to the coördinate planes, *the traces of the same name are always parallel.*

This is in accordance with a theorem of Geometry which proves that when two parallel planes are cut by a third plane the lines of intersection, i.e., the traces, are parallel.

90. The converse of this proposition is also true; *i. e., when the traces of the same name are parallel, the planes in space are parallel.*

An exception arises in the case of traces parallel to *GL*.

91. With the intersecting system of planes, the lines of intersection may assume every conceivable position as the planes themselves alter their position to the coördinate planes.

92. PROBLEM.—*To pass through a given point a plane parallel to a given plane.*

Let (*VP, HP*) be the given plane, and (*p', p''*) the given point (Fig. 84).

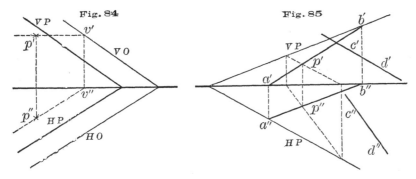

Fig. 84   Fig. 85

Pass through the given point any line parallel to the given plane. The line (*p''v''*, *p'v'*) parallel to a horizontal of that plane is such a line. Find the vertical piercing-point (*v'*, *v''*), and draw the traces *VO, HO* parallel to *VP, HP*, respectively.

93. PROBLEM.—*Through a given line to pass a plane parallel to a second line.*

Let (*a'b'*, *a''b''*) and (*c'd'*, *c''d''*) be the given lines (Fig. 85).

Through any point (*p'*, *p''*) of the first line draw a parallel to the second line. The required plane contains these two lines, and hence the determination of their piercing-points fixes the traces *VP, HP*.

94. PROBLEM.—*Through a given point to pass a plane parallel to two given lines.*

Through the given point pass parallels to the given lines, find their piercing-points, and draw the traces which pass through them.

INTERSECTING PLANES.

95. The intersection of any two surfaces is determined, in general, by the aid of auxiliary secant planes, which pass through the surfaces and cut lines upon them. The points common to the lines thus cut are common to both surfaces and, hence, to their line of intersection.

When the intersecting surfaces are planes, the intersection is a right line, for the determination of which two auxiliary secant planes will ordinarily prove sufficient.

96. PROBLEM.—*To find the line of intersection between two planes, given by their traces.*

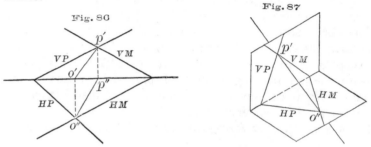

Let (VP, HP) and (VM, HM) be the given traces (Fig. 86).

By the application of the preceding principles, the two coördinate planes may be considered as the auxiliary secant planes. Thus, V cuts the two given planes in the vertical traces, which intersect each other in (p', p''), while H cuts them in the traces HP and HM, which intersect in (o', o''). But the points thus determined are the piercing-points

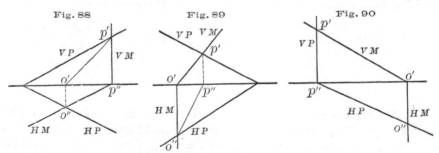

of the line of intersection sought (Fig. 87), the projections of which may be found by Art. 46.

Figs. 88–93 are applications to special cases.

In Fig. 88 the second plane is vertical in position.

In Fig. 89 it is perpendicular to V.

In Fig. 90 the first plane is vertical, the second perpendicular to V.

In Fig. 91 the second plane is perpendicular to *GL*.

In Fig. 92 the line of intersection lies in a plane perpendicular to *GL*.

In Fig. 93 the line of intersection pierces — *H*.

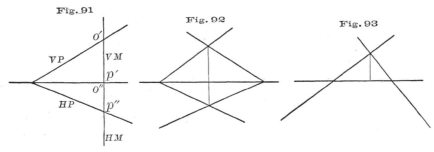

In Fig. 94 the vertical traces are parallel. In this case the line of intersection gives but one piercing-point (*p″*, *p′*) at the intersection of the horizontal traces, and hence is a vertical of each of the given planes.

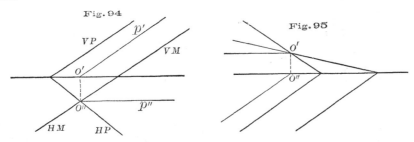

The vertical projection *o′p′* is therefore parallel to the vertical traces, and the horizontal projection *o″p″* is parallel to *GL* (Art. 82).

In Fig. 95 the line of intersection is a horizontal, there being but one piercing-point (*o′*, *o″*) on *V*.

97. In Figs. 96, 97, the traces of the two planes intersect in a common point (*a′*, *a″*) in *GL*. The intersection of the given planes by the

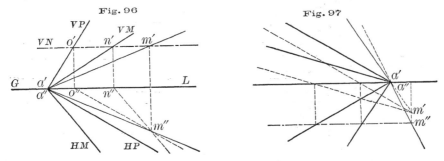

coördinate planes determines this point alone; hence, an additional auxiliary plane must be employed to obtain a second point of the line of intersection.

In Fig. 96 the additional secant plane is horizontal, giving a vertical trace *VN*, and cutting the plane *P* in the horizontal ($o'm'$, $o''m''$), and the plane *M* in the horizontal ($n'm'$, $n''m''$). As these two horizontals lie in the same plane, they intersect in the point (m', m''), and hence ($a'm'$, $a''m''$) is the line of intersection required.

In Fig. 97 the additional secant plane is vertical in position, and cuts the given planes in verticals which intersect in the point (m', m'').

In the first case the line of intersection lies in the first and third dihedral angles; in the second, in the second and fourth dihedral angles.

In Fig. 98 the auxiliary secant plane is any plane *N* assumed within the limits of the drawing. The intersections of this plane with the given planes having been determined as in Fig. 86, the point in which these lines intersect is a point in the line of intersection sought.

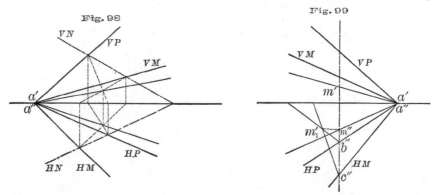

In Fig. 99 the auxiliary secant plane is a new vertical plane perpendicular to *GL*, giving the new vertical traces $b''m_1'$ and $c''m_1'$. Restoring the point m_1' to the primitive planes of projection, ($a'm'$, $a''m''$) is the line of intersection sought.

98. In Figs. 100 and 101 the two intersecting planes are parallel to *GL*; hence, when cut by the coördinate planes, give lines which intersect at infinity. The auxiliary plane may be any plane not parallel to *GL*.

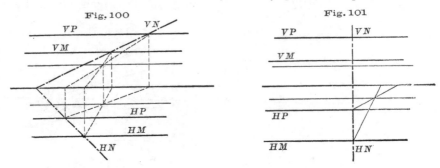

99. In Figs. 102 and 103 the traces do not intersect within the limits of the drawing; hence two points of the line of intersection must be determined by means of two additional auxiliary planes.

In Fig. 102 a horizontal and a vertical secant plane have been employed, giving, respectively, the points (o', o'') and (p', p'') through which the line of intersection passes.

In Fig. 103 the secant planes L and N are parallel to GL, having a common horizontal trace HL and cutting each upon the given planes

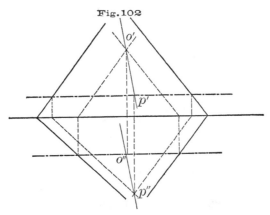

Fig. 102

a line the vertical projections of which are $a'b'$, $c'b'$, $e'd'$, $f'd'$. The points g' and h' in which they intersect determine the vertical projection $g'h'$ of the required line of intersection.

Two additional secant planes, having a vertical trace VQ in common, give in a similar manner the horizontal projections of two lines whose points of intersection determine the horizontal projection of the line of intersection.

100. When the given planes are three in number, the lines of intersection must either be parallel or intersect in a common point.

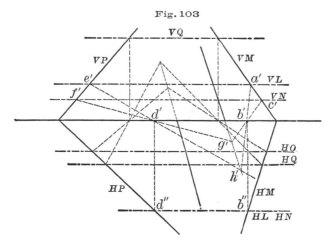

Fig. 103

Their relative positions may be reduced to the following five:

(1) Parallel to one another, giving no intersection.

(2) Two parallel, intersected by the third; giving parallel lines of intersection.

(3) Passing through a common line which becomes their line of intersection.

(4) Intersecting two and two and forming a prismatic surface; giving parallel lines of intersection.

(5) Forming a trihedral angle; giving lines of intersection which meet in a common point.

II. THE LINE AND PLANE.

101. PROBLEM.—*To find the piercing-point of a line on a given plane.*

The solution consists in passing through the given line any auxiliary plane, in determining the line in which this plane cuts the given plane, and in finding the point in which the given line intersects the line thus determined.

Let (*VP, HP*) be the given plane (Figs. 104, 105), and (*a'b', a''b''*) the given line.

The construction becomes extremely simple when the auxiliary secant

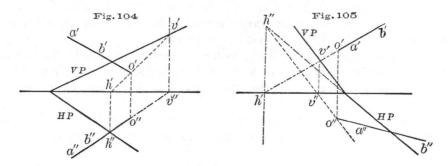

plane passing through the line is assumed to be the projecting plane of the line itself,—in Fig. 104 to *H*, and in Fig. 105 to *V* (Art. 77). Find the line of intersection between the given plane and the projecting plane,

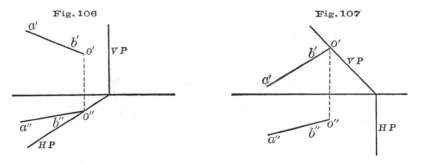

and the point (*o', o''*), in which the line *ab* in space intersects it, is its piercing-point on the given plane.

102. In Figs. 106 and 107 the given plane is perpendicular to either plane of projection.

In Figs. 108 and 109 the plane is given by three points, *a*, *b*, *c*, and the line by its projections ($d'e'$, $d''e''$).

Connect the points by auxiliary lines, and pass through the given line a projecting plane (Art. 101) to *H*, as in Fig. 108, or to *V*, as in Fig. 109, cutting the connecting lines in points which determine the line of

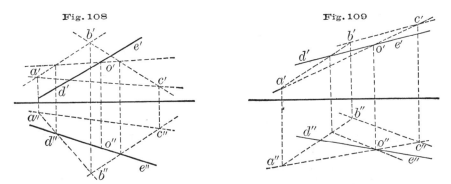

Fig. 108   Fig. 109

intersection. The point (o', o'') in which this line intersects the given line is the piercing-point required.

In Figs. 110 and 111 the given line *ab* lies in a profile plane. In the first case the auxiliary plane has been passed through the given line ($a'b'$, $a''b''$) parallel to *GL*. Its intersection with the given plane *P* is found by the use of additional auxiliary planes.

Thus, through (a', a'') conceive a horizontal secant plane to be passed; it cuts the plane *P* in the horizontal ($c'e'$, $c''e''$), and the auxiliary plane

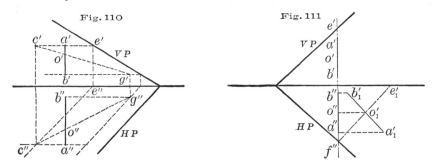

Fig. 110   Fig. 111

in a line ($c'a'$, $c''a''$) parallel to *GL*. The point (c', c'') in which these two lines intersect is a point in the line of intersection sought.

A second point (g', g'') may be determined in a similar way by means of a horizontal secant plane through (b', b''); the intersection (o', o'') of *ab* with the line ($c'g'$, $c''g''$) connecting these two points is the piercing-point required.

In Fig. 111 the auxiliary plane is the profile plane of the given line. By a change of ground-line its intersection $f''e_1'$ with the plane *P*, and the new vertical projection $a_1'b_1'$, may be found, and with them o_1', the supplementary projection of the piercing-point required.

103. THEOREM.—*A right line perpendicular to a plane in space gives projections which are respectively perpendicular to the traces.*

In Fig. 112 let P be the given plane, and ab a line perpendicular to it. The plane projecting ab upon H is not only perpendicular to that plane, but also to the given plane P; hence, being perpendicular to two planes, it is, by Geometry, perpendicular to their line of intersection HP, or the horizontal trace. From this it follows that HP is perpendicular to every line in the projecting plane which passes through its foot on that plane, and thus to $a''b''$, the horizontal projection of the given line.

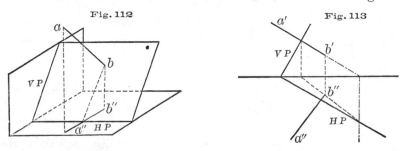

Fig. 112   Fig. 113

The converse of this proposition is likewise true, viz.: *if the projections of a line are perpendicular respectively to the traces of a plane, the line is perpendicular to the plane.*

Fig. 113 represents the graphical solution of the case.

104. PROBLEM.—*Through a given point in space to pass a plane perpendicular to a given line.*

Let (a', a'') be the given point (Fig. 114), and $(b'c', b''c'')$ the given line.

By the conditions of the problem the traces of the required plane must be respectively perpendicular to the projections of the given line. Through a lead a vertical $(a'd', a''d'')$ of that plane, find its piercing-point (d', d''), and draw the traces perpendicular to the projections of the given line.

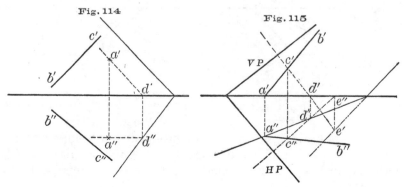

Fig. 114   Fig. 115

105. PROBLEM.—*Through a given line to pass a plane perpendicular to a given plane.*

Let $(a'b', a''b'')$ be the given line, and (VP, HP) the given plane (Fig. 115).

From any point, as (c', c''), of the given line lead a perpendicular $(c'd', c''d'')$ to the plane, find the piercing-points (a', a''), (d', d'') and (e', e'') of these two lines, and the traces of the required plane pass through them.

106. PROBLEM.—*To pass a plane through a given point and parallel to two given lines.*

Let (a', a'') be the given point, and $(b'c', b''c'')$, $(d'e', d''e'')$ the given lines (Fig. 116).

Through a lead lines parallel to the given lines and find their piercing-points; the traces of the plane pass through them.

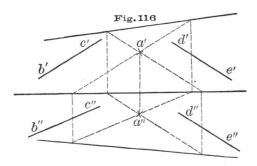

Fig. 116

107. PROBLEM.—*To project a given point upon a plane.*

Through the point pass a line perpendicular to the plane (Art. 103), and find its piercing-point upon it.

108. PROBLEM.—*To pass through a given point a line which intersects two given right lines.*

1st Solution.—Through each line and the given point pass a plane; their line of intersection is the line sought.

2d Solution.—Determine the traces of a plane which contains one of the given lines and the point, find the piercing-point of the second line upon this plane, and connect the point so found with the given point.

CHAPTER V.

I. THE GEOMETRICAL SURFACES.

109. Every surface may be regarded as generated by the movement of a line either fixed or variable in form, and regulated in its change of position in accordance with some definite law.

Thus, a plane may be described by the motion of a right line which, constantly touching a second right line, remains parallel to its original position; or, again, it may be described by turning the first line around the second in such a way that the two shall remain at right angles to each other and touch in a fixed point.

The moving line is termed the *generatrix*, and the line which guides the movement the *directrix*.

101. Surfaces may be divided into two general classes:

(1) The *ruled* or *rectilinear*, or such as may be generated by the motion of a right line.

(2) The *curvilinear* proper, or such as admit of no rectilinear generatrices.

The following table exhibits these classes and their subdivisions:

SURFACES $\begin{cases} \text{RULED} \begin{cases} \textit{Plane} \\ \textit{Single-curved} \begin{cases} \text{Developable} \\ \text{Non-Developable or Warped} \end{cases} \end{cases} \\ \text{DOUBLE-CURVED} \end{cases}$

111. The *plane* surfaces are such as are limited by plane faces, as prisms, pyramids, etc.

The *single-curved* surfaces are those which admit of rectilinear generatrices, and are of three kinds:

(1) *Cylindrical* surfaces, in which all the positions of the generatrix are parallel.

(2) *Conical* surfaces, in which all the positions of the generatrix *intersect* in a common point.

(3) Surfaces in which the rectilinear elements intersect two and two.

The *double-curved* surfaces are those on which no right line can be drawn, as spheres, ellipsoids, etc., and which, therefore, can only be generated by curves.

112. Of the single-curved surfaces the *developable* are those in which successive rectilinear elements can be brought in contact with a plane without crumpling, folding or tearing; the *warped* are those in which such a procedure is, in general, impossible.

II. PLANE SURFACES.

113. The *prism* is a surface generated by the motion of a right line which, while remaining parallel to its original position, glides along the perimeter of any polygon.

It may likewise be generated by a polygon which, while remaining parallel to its original position, has one of its points in constant contact, during its motion, with a given line. Thus, the directrix of the first generation may become the generatrix of the second, and reciprocally.

The surface thus determined, if limited in extension by planes, has two bases and an altitude equal to the perpendicular distance between the two.

When the base is a parallelogram, the prism becomes the *parallelopiped;* when all the faces are squares, the *cube.*

114. A *pyramid* is a surface generated by the right line, one point of which remains fixed while the line itself glides around the perimeter of any polygon.

Should the generatrix extend beyond the fixed point, termed the *vertex,* an upper and a lower surface will be formed, the limited portions of which are determined by the *bases.*

The lines drawn from the vertex to the angular points of the base and marking the intersections of the adjacent faces are termed the *edges.*

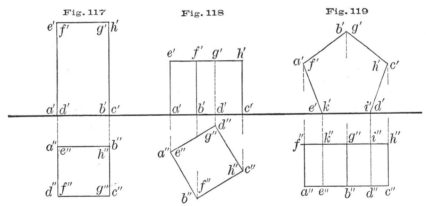

Fig. 117   Fig. 118   Fig. 119

In the prism the edges are parallel; in the pyramid they converge to a common point, the vertex.

Both surfaces take their distinctive names from their bases, as triangular, hexagonal, etc.

115. PROBLEM.—*To find the projections of a parallelopiped of given dimensions.*

Let the base be assumed as a square of 1 inch, the altitude 2 inches, and let the prism stand on *H* at a distance of 1 inch from *V.*

(1) The lower base resting on *H* is its own projection (Fig. 117) on that plane; hence, draw *a″b″c″d″* a square of the given dimensions, and mark its vertical projection on *GL.*

(2) The axis being vertical, so are the edges; hence, measure above *GL* a distance equal to the altitude, or 2 inches, and draw a horizontal iine.

(3) The bases being equal and parallel, their projections will respectively follow the same conditions; hence, *e'f'g'h'* is the vertical projection of the upper base, and *a'e'*, *d'f'*, *c'g'*, *b'h'* the projections of the edges.

116. Fig. 118 represents the projections of a cube whose faces incline to *V* and are perpendicular to *H*.

Fig. 119 represents the projections of a right pentagonal prism whose bases are parallel to *V*, and whose edges, as a consequence, are parallel to *H*.

117. PROBLEM.—*To project a right pyramid of given dimensions.*

Let the base be a hexagon of 1½ inches, the altitude 3 inches, and the pyramid stand on *H*, 1 inch distant from *V*.

(1) Measure 1 inch below *GL*, and draw the side *a''b''* of the hexagon equal to 1½ inches. Construct the hexagon on this line as a base, and find its vertical projection in *GL* (Fig. 120).

(2) The axis being vertical and passing through the centre of the base, *v''* is its horizontal projection, and *v'v'*, equal to 3 inches, its vertical projection.

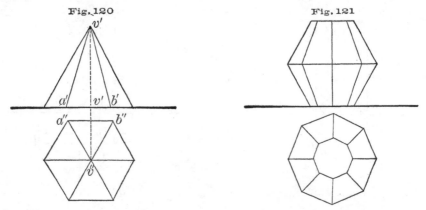

Fig. 120   Fig. 121

(3) The upper extremity of the axis is the vertex; hence the edges are determined by joining its projections (*v'*, *v''*) with those of the angular points of the base.

Fig. 121 represents the projections of the double octagonal pyramid from which the solid angles have been cut by planes parallel to the common base.

118. Among the plane surfaces are the regular polyhedrons, or those in which the solid angles are equal. They are five in number, viz.: the *tetrahedron*, in which the solid angles are bounded by three equilateral triangles; the *cube*, in which they are bounded by three squares; the *octahedron*, by four equilateral triangles; the *dodecahedron*, by three pentagons; and the *icosahedron*, by five equilateral triangles.

119. PROBLEM.—*To determine the projections of the octahedron.*

Assume that one face lies in H with one side parallel to GL (Fig. 122).

On the given side ($a'b'$, $a''b''$) construct an equilateral triangle. As a square, *abde*, whose sides form the common base of the two pyramids lies in a plane parallel to GL, its projection on a new vertical plane G_1L_1 is a line equal to the given side, and forming at its middle point a right angle with the projection of the axis on that plane. Hence from a_1' as a centre describe an arc with a radius equal to one half of the given side; the line $c_1'f_1'$ is the new vertical projection of the axis, and $a_1'b_1'd_1'e_1'$ that of the base, from which $a''b''d''e''$ may be readily determined.

As the faces diagonally opposite are parallel to each other, the uppermost one is parallel to H; hence ($d''e''f''$) and ($d'e'f'$) are its horizontal and vertical projections respectively. Join the vertices (c', c'') (f', f'') with the angular points of the base and complete the surface.

Fig. 122

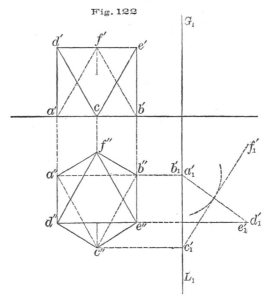

120. PROBLEM.—*To determine the projections of the icosahedron.*

Assume the axis of the surface to be vertical (Fig. 123).

On a given side ($a''b''$, $a'b'$) construct a regular pentagon ($a''b''c''d''e''$, $a'b'c'd'e'$), having one radius parallel to GL.

As the upper portion of the surface is a regular pyramid, joining m'' with the angular points of the pentagon determines its horizontal projection. To determine the vertical projection make $d'm'$ equal to the given side, since this edge of the pyramid is parallel to V.

Construct a second pentagon, equal to and concentric with the first, but in such a way that its angular points $f''g''h''i''k''$ shall lie in the middle of the arcs which are subtended by the sides of the first pentagon. Join the vertex (n'', n') with each of these points to form the lower pyramid of the surface, and from each angular point of the upper pentagon draw lines to the two adjacent points of the lower pentagon, thus determining the intermediate zone between the two pyramids.

As $c''h''$ is by this construction parallel to V, the vertical projection $c'h'$ is shown in its true length; hence, the plane of the lower pentagon is fixed by making the distance from c' to h' equal to the given side.

121. As polyhedra are represented by the projections of the edges which limit the faces, and as, by the preceding examples, it is manifest that all the faces are not visible, it will be well, just here, to consider in

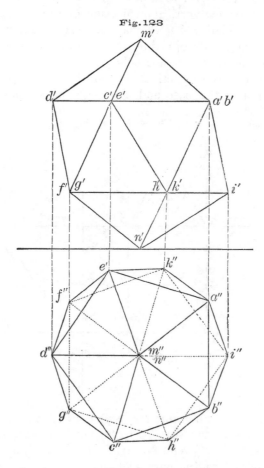

Fig. 123

what way the projections are affected and the means by which they may be represented.

The point at which the eye is located in space is the point of sight. If its distance from the object be finite, the lines—termed visual rays— drawn from the eye to each point of that object will be divergent, and the projections, though similar to the original in respect of form and position, will be larger or smaller as the object is placed before or behind the plane of projection (Fig. 124).

In orthographic projection, if the projecting lines be regarded as visual rays (Fig. 125), the point of sight must be at an infinite distance in order that their parallelism may be established. Whence it follows that the horizontal projection is that view of an object which is obtained at an infinite distance from the horizontal plane, and the vertical projection is that view of an object which is obtained at an infinite distance from the vertical plane.

122. In the vertical projection, therefore, those parts of the same object or of any other object which lie to the front may cover or conceal those which lie to the rear; and, in a similar way, the upper parts of the same or any other object may, in the horizontal projection, cover or conceal the lower parts. The solution of these difficulties will be materially simplified by the observance of the following facts:

(1) The outlines or extreme limiting lines of both projections are *always visible*.

(2) Only that part of the vertical projection can possibly be concealed which is covered by some other part in the same projection. The horizontal projection of the parts *vertically in common* will determine the question.

(3) Only that part of the horizontal projection can possibly be concealed which is covered by some other part in the same projection. The

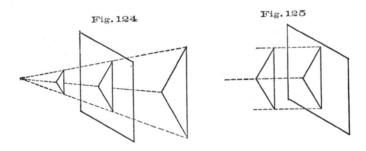

Fig. 124   Fig. 125

vertical projection of the parts *horizontally in common* will determine the question.

(4) In passing from a part seen to a part concealed, the outline of the projection must be crossed.

123. Figs. 126 and 127 will serve as illustrations.

(1) In Fig. 126 the outline of the *combined* projections of the three solids has been drawn in full lines, in accordance with the first principle.

(2) An inspection of the horizontal projection shows that a part of each surface is included within the limits of this outline. It is manifest, however, from an examination of these same parts in the vertical projection, that the upper base of the triangular prism rises above the other two surfaces; hence, in looking down upon H, the prism will conceal such parts of them as lie within its contour. In like manner it conceals one side of the lower base ($a'b'$, $a''b''$) and the under face ($a'b'c'd'$, $a''b''c''d''$) (Fig. 127).

(3) An inspection of the vertical projection (Fig. 126) shows that a

part of each surface is included within the general contour; but by an examination of the horizontal projection it is evident that the cylinder lies to the front of the other two surfaces, and hence covers and conceals all those parts of the other surfaces which lie within its limits. It will also be seen that the pentagonal prism covers a part of the triangular prism.

The dotted lines indicate the hidden parts.

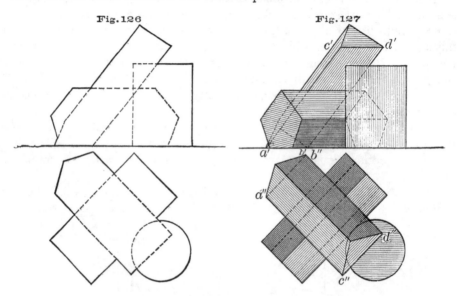

124. When the methods of Descriptive Geometry are applied to machinery, architecture, etc., the term *plan* is employed to denote the horizontal projection, and *elevation* the vertical projection; *front elevation* referring to an ordinary vertical projection, and *side elevation* to that which is determined by means of a profile plane.

125. PROBLEM.—*To project an oblique pyramid of given dimensions.*

Let the pyramid (Fig. 128) be pentagonal, the radius of circumscribing circle $\frac{3}{4}$ of an inch, the altitude 2 inches, the axis inclining to the base at an angle of 45° to the left; and let it stand on *H*, 1 inch from *V*.

(1) Place the surface in its simplest position with its axis parallel to *V*. Find the projections of the base (Art. 117).

(2) The axis being parallel to *V*, find its projection on that plane, and determine its horizontal projection parallel to *GL*.

(3) Join the upper extremity or vertex (*v'*, *v''*) with the respective angular points of the base, observing that for the vertical projection the edge (*a'v'*, *a''v''*) at the back of the surface is concealed, and for the horizontal projection the edge (*b'v'*, *b''v''*) and two sides of the base (*a''b''*, *a'b'*) (*b''c''*, *b'c'*) on the under part of the surface are concealed.

126. PROBLEM.—*To project an oblique prism of given dimensions.*

Let the dimensions be as in the preceding case (Fig. 129), but the axis parallel to *H* and inclining to *V* towards the right hand.

(1) Place the prism with its bases parallel to V, and determine the vertical projection of the one nearer to that plane; $a''d''$ is its horizontal projection.

(2) The axis being parallel to H, determine its horizontal projection $o''o''$ of the given length and in the required position, $o'o'$ is its vertical projection.

(3) The two bases being parallel, the projections are likewise parallel; hence the vertical projection of the second base has its homologous

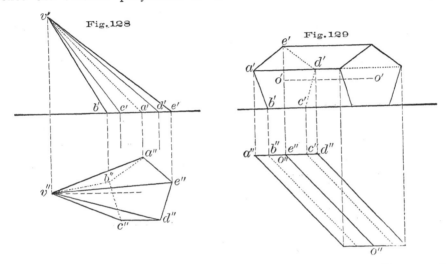

sides parallel to the first, and its horizontal projection a line parallel to *GL*.

(4) Join the corresponding angular points of the bases to determine the edges.

III. SINGLE-CURVED SURFACES.

127. The *cylinder* is a surface generated by the motion of a right line which constantly touches the perimeter of any curve and remains parallel to its original position. The different positions of the generatrix indicate the *rectilinear elements* of the surface.

The surface may also be generated by the motion of a curve in such a way that each point shall describe a line parallel to a given right line. The different positions of the curve indicate the *curvilinear elements*. The same distinctions as to bases, axes, etc., are to be noticed as in the case of the prism, an analogy at once apparent, as the curve, whether regarded as generatrix or directrix, may be considered as a polygon of an infinite number of sides.

Cylinders are named according to the character of the base, as circular, elliptical, parabolic, etc.

128. The *cone* is a surface generated by the motion of a right line which passes through a *fixed* point, the vertex, and constantly touches the perimeter of any curve. If the generatrix extend beyond the fixed point it will describe two branches of the surface, one on either side of the vertex, termed the *upper* and *lower nappes*. The different positions of the generatrix indicate the *rectilinear elements* of the surface.

The cone may also be generated by the motion of a curve in such a way that any point in its plane shall move along a right line passing through a fixed point, and the curve vary in proportion to its distance from that point. The different positions of the curve indicate the *curvilinear elements* of the surface.

If the vertex of the cone be assumed at a maximum distance, the surface becomes cylindrical; if at a minimum distance, a plane. If the radius be infinitely increased, the surface becomes a plane; if infinitely diminished, a right line.

129. PROBLEM.—*To project a right cone of given dimensions.*

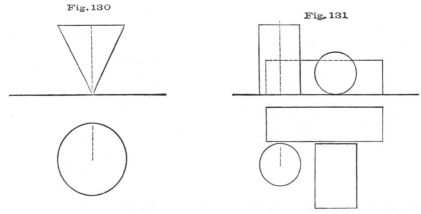

Fig. 130   Fig. 131

The solution is precisely similar to that of the right pyramid, the base of the cone being regarded as a polygon of an infinite number of sides; but as the edges now become the rectilinear elements of the surface, only those appear in the projections (Fig. 130) which lie in the extreme projecting planes to either plane of projection. The same holds true of the right cylirder (Fig. 131).

130. PROBLEM.—*To project an oblique cone of given dimensions.*

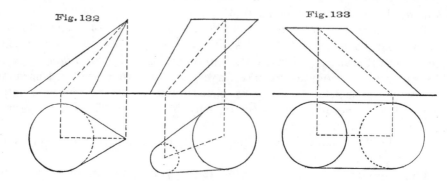

Fig. 132   Fig. 133

The solution is similar to that of the oblique pyramid, with this difference: that, there being no distinctive elements or edges on the surface, those only are represented which form the extremes or tangents to the projections (Fig. 132).

The same holds true of the oblique cylinder (Fig. 133).

131. The *warped* surfaces are those which are generated by the right line, but cannot be brought in contact or spread upon a plane without folding or tearing.

The simplest of these surfaces is the *hyperboloid of one nappe,* which is generated by the motion of a right line which constantly touches three directrices (Fig. 134).

Thus, let ($a'b'$, $a''b''$) be a given right line and ($c'd'$, $c''d''$) a vertical axis around which the given line is to revolve in such a way that its two extremities and the intermediate point (e', e'') shall glide upon the perimeters of three given circles. By these conditions the rectilinear generatrix preserves the same relative position to and distance from the

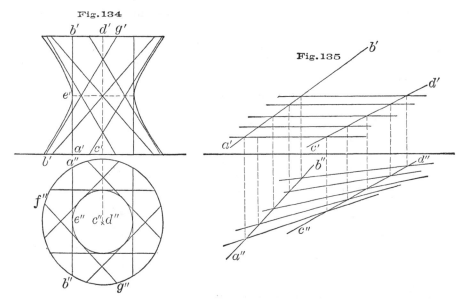

axis; hence its horizontal projections $a''b''$, $f''g''$, etc., must be tangent to the smallest circle, the circle of the *gorge,* on which the point e glides. As the upper and lower extremities lie in the larger circles, the vertical projections are readily determined by means of ordinates.

132. The *hyperbolic-paraboloid* or *warped plane* is a surface (Fig. 135) which is generated by the motion of a right line which glides upon two right lines not in the same plane, and remains constantly parallel to a given plane, termed the *plane directer.*

Thus, let ($a'b'$, $a''b''$) and ($c'd'$, $c''d''$) be the two right directrices, and let H be the plane directer. Any plane parallel to H cuts the two lines in points which determine the position of the generatrix.

133. The *right conoid* is a surface generated by a right line which glides along a curve in such a way that it constantly touches a right line and remains parallel to a plane directer.

Thus, let $(a'e'c', a''e''c'')$ be the given curve (Fig. 136), and $(b'd', b''d'')$ be the given line, and let H be the plane directer.

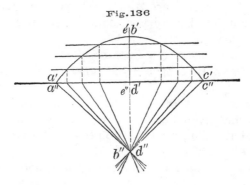

Fig. 136

Any horizontal plane cuts the curve and line in points which determine the generatrix in position.

134. The *helicoids* are surfaces generated by the right line which glides along a helix, and maintains an invariable position to the axis of the curve.

The *helix* is a curve generated by a point which moves along the

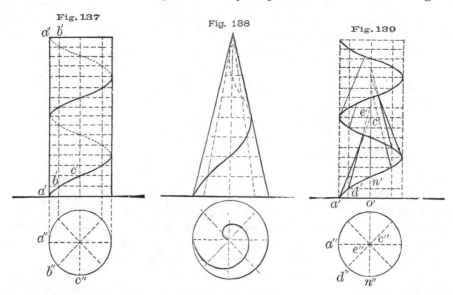

Fig. 137   Fig. 138   Fig. 139

surface of a cylinder in such a way that a constant ratio is maintained between the measure of its rotation and ascent.

Thus, let $(a'a', a'')$ be the element of the cylinder, and let its foot be the moving point (Fig. 137). Divide the circumference of the base into eight and the height into sixteen equal parts, and let it be assumed that for every turn of an eighth around the axis the point (a', a'') ascends one sixteenth of the height. Then when the element has moved into the position ($b'b'$, b'') the point will have risen to (b', b''), and when the third position ($c'c'$, c'') has been attained to (c', c''), and so on in regular succession until two complete turns have been made.

135. PROBLEM.—*To project a warped helicoid.*

There are two surfaces belonging to this class which are to be noted: one in which the generating line intersects the axis at right angles, and another in which it intersects it at an invariable acute angle.

Let the second case be assumed (Fig. 139), and let ($a'c'$, $a''c''$) be the initial position of the generatrix. As the generatrices are all equally inclined to the axis, the distance between their extremities for any position will be constant; hence, to determine a second generatrix ($d'e'$, $d''e''$) for a second point (d', d'') of the helix, lay off on the axis the distance $n'e'$ equal to $o'c'$. Proceed in a similar manner for the remaining positions of the generatrices.

IV. DOUBLE-CURVED SURFACES AND SURFACES OF REVOLUTION,

136. A surface of revolution is generated by any line which revolves around a right line, termed the *axis of rotation.*

When the generatrix is plane it is termed a *meridian line,* and its plane a *meridian plane.* During rotation every point of the meridian describes the circumference of a circle whose plane is perpendicular to the axis and whose centre lies in it. These circles are the *parallels,* and may also be generatrices of the surface.

137. While all the meridians are equal, the parallels vary in size according to the nature of the generating line. When the meridian becomes a right line, the surface is either the right cylinder or cone as the generatrix is parallel to or intersects the axis.

138. Among the geometrical surfaces of revolution the following are to be noted:

(1) *Ellipsoid* of revolution, which is generated by the complete revolution of a semi-ellipse around its major or minor axis, giving, respectively, the *prolate* and *oblate* ellipsoids.

(2) *Hyperboloid* of revolution, which is generated by the complete revolution of a semi-hyperbola around one of its axes. It is indefinite in extent, being limited for constructive purposes by planes perpendicular to the axis.

(3) *Paraboloid* of revolution, which is generated by the complete revolution of a semi-parabola around an axis of the curve. It is also indefinite in extent and limited as in the preceding case.

(4) *Sphere,* generated by the semicircle.

139. The projections of surfaces of revolution are, in general, determined by the meridians and parallels.

The case is much simplified should the axis of the surface be taken perpendicular to *H.* In such a position the vertical projection is a full meridian, and the horizontal projection a series of concentric circles

whose number will depend upon the important parallels characteristic of
the surface.

140. In addition to the surfaces of revolution the following are to
be noted :

(1) *Tri-axial ellipsoid* (Fig. 140), which is generated by an ellipse *aba'b'*
which moves in such a way that, while it remains parallel and similar
to itself, its centre *c* glides along the axis *AA'*, and the extremities of
the major axis describe the ellipse *aAa'A'* at the same time that the extremities of the minor axis *b* and *b'* describe the ellipse *bAb'A'*.

When *BD* equals *CD* the ellipsoid becomes one of revolution, and
when *BD = CD = DA* it becomes the sphere.

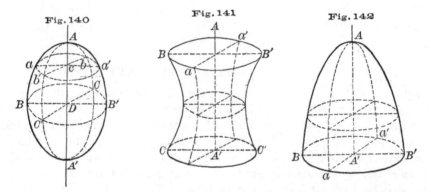

(2) *Hyperboloid of one nappe* (Fig. 141) with three unequal axes, which
is generated by an ellipse *BaB'a'* which moves in such a way that, while
it remains parallel and similar to itself, its centre glides along the axis
AA', and the extremities of either axis have a given hyperbola *BCB'C'*
for a directrix.

When the axes are equal the surface is one of revolution.

(3) *Elliptic paraboloid* (Fig. 142), which is generated by an ellipse *BaB'a'*
which, while remaining parallel and similar to itself, moves so that the
extremities of either axis have the parabola *BAB'* for a directrix.

When the axes are equal the surface is one of revolution.

CHAPTER VI.

SUPPLEMENTARY PLANES AND PROJECTIONS.

141. It frequently happens in practice—and illustrations are not wanting in the preceding articles—that, in addition to the coördinate planes, other ways and means must be devised to render the graphical solution of a problem possible. The change in the position of the ground-line is an expedient of this character. The necessity for such special constructions arises from some special arrangement of the objects in space with reference to the coördinate planes.

142. Under such conditions the operations may be simplified in two ways: either by *means of additional planes of projection or by some new arrangement of the objects themselves.* In other words, the questions to be resolved are:

(1) Having the primitive projections of any object, to determine new or *supplementary* projections on additional planes.

(2) Having the primitive projections of any object, to determine its projections after a change in *its own position* to the primitive coördinate planes.

The discussion of the first condition is the subject-matter of this chapter; the second will be treated in a succeeding chapter.

143. The employment of an additional plane in nowise affects the position of the object to the primitive planes; its sole end is to present some new *view* of that object, thereby admitting of a clearer comprehension of its character and lessening the constructive difficulties of the case.

144. The position of the supplementary plane must mainly depend upon that of the object, although, wherever practicable, it is desirable that it should be as simple as possible; hence, perpendicular to both coördinate planes, or, at least, perpendicular to one of those planes.

145. Whatever position may be adopted for the supplementary plane, the projections are determined precisely as in the case of the primitive planes; that is, by letting fall from the object perpendiculars to that plane, and by marking their intersections thereon.

146. The constructions thus made are brought into the drawing plane by revolving the supplementary plane around either trace into either of the coördinate planes.

Thus with S the supplementary plane perpendicular to both coördinate planes (Fig. 143), it may be revolved:

(1) Around VS, termed the *vertical supplementary ground-line*, into V (Fig. 144).

(2) Around *HS*, the *horizontal supplementary ground-line*, into *H* (Fig. 145).

147. The results of these constructions demonstrate that:

(1) The vertical trace or ground-line *VS* (Fig. 144) remains fixed, while the horizontal ground-line *HS* rotates through a quadrant until it coincides with *GL*.

(2) All distances from *H* are measured on *S* at right angles to *HS*.

(3) All distances from *V* are measured on *S* at right angles to *VS*.

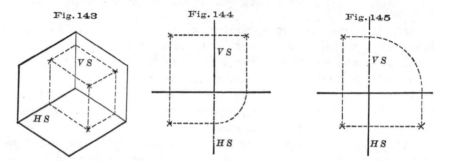

(4) The projecting lines to *S* are represented on the coördinate planes by ordinates perpendicular respectively to the supplementary ground-lines.

148. With *S* perpendicular to either plane of projection and inclining to the other, the supplementary ground-lines assume the positions indicated in Art. 68. In such a case *S* is ordinarily revolved around the inclining ground-line into the primitive plane in which that line lies.

Thus, in Fig. 146 *S* has been revolved around *VS* into *V*, and in Fig. 147 around *HS* into *H*.

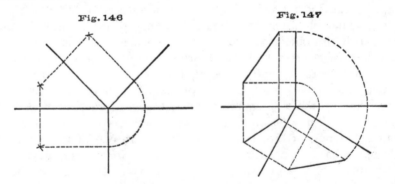

149. The results of this rotation demonstrate that—

(1) The inclining ground-line remains fixed during the rotation.

(2) As the relative position of the supplementary ground-lines is not altered by the rotation, they still retain a rectangular position to each other.

(3) The projecting lines let fall from the object upon *S* are represented by the ordinates drawn from the primitive projections perpendicular to the supplementary ground-lines.

(4) The intersections of these ordinates determine the required supplementary projections.

150. When *S* is made to incline to both coördinate planes—a position least calculated to insure simplicity of construction—the projections of the object will still be found by letting fall perpendicular projecting lines upon that plane. As in the preceding cases, the ordinates must necessarily be perpendicular to the respective ground-lines.

151. The supplementary plane is rotated away from the object in order to avoid the confusion which would necessarily arise from the superimposition of the projections.

152. PROBLEM.—*The traces of two planes being given, to find their line of intersection by means of a profile supplementary plane.*

Let (*VO, HO*) and (*VP, HP*) be the given traces (Fig. 148), and let *S* be revolved into *V*.

The points in which *HO* and *HP* cut the ground-line *HS* are carried with that line into *GL*, while the points in which *VO* and *VP* cut *VS* remain fixed; hence *SO* and *SP* are the supplementary traces, and their point of intersection a_1' the supplementary projection of the line of intersection. As the distances from the supplementary ground-lines measure the distances from the coördinate planes, $a a_1'$ is the height above *H*, and $\beta a_1'$ the distance behind *V*. The intersection is effected in the second angle, and (*a'b', a''b''*) are the projections sought.

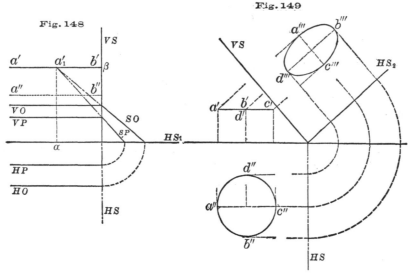

Fig. 148   Fig. 149

153. PROBLEM.—*To determine the oblique supplementary projection of a circle, its primitive projections being given.*

Let (*VS, HS*) be the supplementary plane, and (*a'..b'..d', a''..b''..d''*) the given circle.

From any number of points of the circle (Fig. 149) let fall projecting lines upon *S*, which lines being perpendicular to that plane are projected in ordinates perpendicular respectively to *VS* and *HS*.

The points in which the ordinates cut *HS* are revolved with that line into the new position *HS,*, and determine the several ground-points. The ordinates drawn from these points intersect those drawn from the vertical projection in (*a'''*, *b'''*, *c'''*, *d'''*), through which the required projection of the circle passes.

154. PROBLEM.—*To find the primitive projections of a right prism, its supplementary projection being given.*

Let (*VS*, *HS*) be the given plane (Fig. 150), and (*a'''* ... *h'''*) the supplementary projection of the prism.

Assuming the prism to be right, the axis must be parallel to *GL*, and hence both the edges and their projections. As the distances from *V*

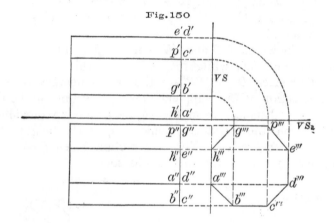

Fig.150

are measured from *VS* in the supplementary plane, or from *VS,* in the revolved position, the horizontal projection may be determined immediately by ordinates drawn from the supplementary projection parallel to *GL*.

And, again, as the distances from *HS* measure the heights above *H*, parallels to *HS* drawn through the points of the supplementary projection mark these distances upon *VS,*; whence, upon restoring *S* to its original position, they are transferred to *VS* and determine the vertical projection.

CHAPTER VII.

CHANGE OF POSITION BY ROTATION AND RABATTEMENT.

155. Instead of employing the method of the preceding chapter the same end may be attained by changing the position of the object itself, thus affording new views and consequently new projections.

The operations by which this is accomplished are either movements parallel to a rectilinear or plane director, or movements of rotation.

156. In these changes it is to be remembered that the principal object is to facilitate the work of solution; hence, whatever the alteration in position, the simplification of the constructions must be kept constantly in view.

157. With the movement of rotation there are necessarily implied:

(1) An *axis* around which the object revolves—the *axis of rotation AB* (Fig. 151).

(2) The fixed distances of each point of the object from the axis during the entire rotation—the *radii of rotation CD*.

(3) The foot of the radius, always marking a point in the axis—the *centre of rotation C*.

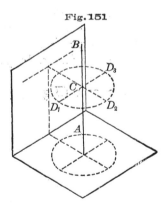

Fig. 151

(4) The *locus* of each point—the circle of rotation $D_1 D_2 D_3$.

(5) The plane of that circle, always at right angles to the axis—the *plane of rotation*.

(6) The *arc* $D_1 D_2$ through which any point D_1 revolves, giving the *measure of rotation* $D_1 C D_2$ for every other point of the object.

Taken together, these constitute a system of rotation the position of which must evidently depend upon that of the axis.

158. There are two general positions which such an axis may assume to the moving object: it may have one or more points in common with it, or may lie wholly outside it.

In connection with the first case the more ordinary positions of the axis are:

(1) With a line, it has a point in common.

(2) With a plane, a point or line in common.

(3) With a plane figure, it coincides with an axis, diameter or side, or with any tangent to it.

(4) With a surface, it coincides with an axis, element or tangent.

159. In the second case the ordinary positions of the axis are:

(1) With a line, it is either parallel to it or lies in another plane.

(2) With a plane, it is parallel to it.

(3) With a plane figure it lies in the plane of the figure or is parallel to it.

(4) With a surface, it is parallel to some line or element.

160. While the axis of rotation may be made to assume any position to the planes of projection, still for practical purposes it ought to be so placed as to render the constructions as simple as possible. Such a position is one in which the axis is assumed to be perpendicular to either plane of projection.

I. ROTATION OF THE POINT.

161. In order to effect the rotation of a point around an axis, a perpendicular must always be passed through the point to the axis, giving the centre of rotation. This perpendicular, which is the line of the radius, must be turned into the required position and the original distance of the point from the axis set off upon it, the measurement being made from the foot of the perpendicular or centre.

162. PROBLEM.—*To revolve a point into either plane of projection around an axis lying in that plane.*

Let $(x''y'', x'y')$ be the given axis (Fig. 152), and (a', a'') the given

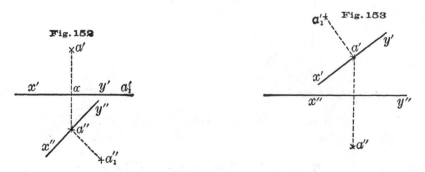

point. The perpendicular drawn from the point in space to the axis is its projecting line to H, the plane in which that axis lies; hence, a'' is the centre of rotation, and aa' the true length of the radius. Through

a'' draw an indefinite perpendicular to the axis, and lay off a distance $a''a_1''$ equal to $\alpha a'$; (a_1'', a_1') are the projections of the point in the revolved position. In Fig. 153 the axis is a line of the vertical plane.

163. PROBLEM.—*To revolve a point into either plane of projection around an axis lying in that plane when neither projection of the point coincides with the projection of the axis.*

It will be seen from Fig. 154 that the perpendicular ac'' let fall from

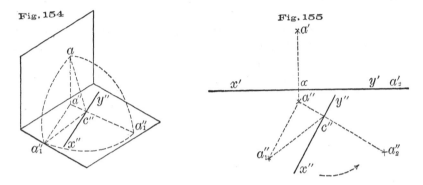

the point a to the axis $x''y''$ is the hypothenuse of a triangle $ac''a''$, right-angled at a'', in which aa'', the perpendicular, is equal to the distance of the point from H, and $a''c''$, the base, is equal to the distance of a'' from the axis.

Hence, in Fig. 155, (a', a'') being the given point and $(x''y'', x'y')$ the axis, let fall from a'' an indefinite perpendicular to $x''y''$, and mark the centre of rotation (c'', c'); then $a''c''$ is the base of a triangle of which

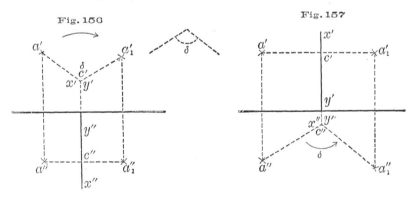

aa', the height above H, is the perpendicular. At a'' erect a perpendicular to $a''c''$, and lay off a distance $a''a_1''$ equal to $\alpha a'$, then with c'' as a centre and a radius equal to $c''a_1''$ describe an arc until it intersects $a''c''$ prolonged; (a_2'', a_2') is the revolved point sought.

164. PROBLEM.—*To revolve a point through a given angle around a horizontal axis.*

Let (a', a'') be the given point, and $(x'y', x''y'')$ the axis (Fig. 156), and let δ be the given angle. As the plane of rotation is parallel to V, the radius drawn from the point to the axis is likewise parallel to V during its entire rotation; hence, $a'c'$ measures the true length of the radius, and $a''c''$ is its horizontal projection. Draw $c'a_1'$, the revolved position of the radius, so that $a'c'a_1'$ measures the required angle; (a_1', a_1'') is the revolved point sought.

165. PROBLEM.—*To revolve a point through a given angle around a vertical axis.*

Let (a', a'') be the given point, $(x'y', x''y'')$ the axis, and δ the angle (Fig. 157).

As the plane of rotation is parallel to H, the radius drawn from the point to the axis is likewise parallel to H during its entire rotation; hence, $a''c''$ measures the true length of the radius, and $a'c'$ is its vertical projection. Draw $c''a_1''$, the revolved position of the radius, so that $a''c''a_1''$ measures the required angle; (a_1'', a_1') is the revolved point sought.

166. PROBLEM.—*To revolve a point through a given angle and around any axis parallel to the horizontal plane of projection.*

Let (a', a'') be the given point, $(x'y', x''y'')$ the axis, and δ the angle (Fig. 158).

The plane of the circle of rotation, being at right angles with the axis, is parallel to neither coördinate plane; hence the true size of the triangle,

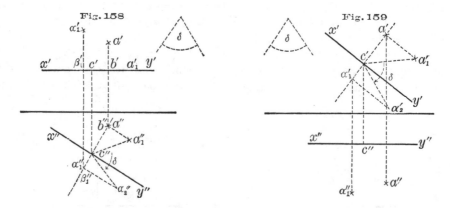

Fig. 158   Fig. 159

the hypothenuse of which determines the length of the radius, must be found. This is effected by bringing the plane of rotation parallel to either coördinate plane; in this case preferably to H.

The perpendicular let fall from the given point (a', a'') to the axis lies in a plane perpendicular to it, giving a horizontal trace $(a''c'')$ perpendicular to $x''y''$ and cutting the axis in the point (c'', c'). The distance of this point from the given point is the radius of rotation, being the hypothenuse of the right-angled triangle whose base is equal to $c''b''$ and whose perpendicular is equal to the distance $a'b'$ of the given point above the horizontal plane of the axis.

Turn this triangle around its base $(b''c'', b'c')$ until it is brought parallel to H; the base remains fixed while the given point is projected at (a_1'', a_1'), the distance $a''a_1''$ being equal to $a'b'$; hence $a_1''c''$ is the hypothenuse or true length of the radius let fall from the point a upon the axis xy in space. From c'' as a centre and with this radius describe an arc whose measure is the given angle δ and determine α_2''.

By a counter-rotation restore the plane of the triangle to its original position, carrying with it the line $c''\alpha_2''$, the hypothenuse of a second triangle whose base, $c''\alpha_1''$, determines α_1'', the horizontal projection of the revolved point, and whose perpendicular, $\alpha_1''\alpha_2''$, measured above β_1' determines α_1', the vertical projection of that point.

167. PROBLEM.—*To revolve a point through a given angle around any axis parallel to the vertical plane of projection.*

Fig. 159 shows · constructions which are similar in every respect to those of the preceding case.

168. PROBLEM.—*To revolve a point through a given angle around any axis.*

Assume a new ground-line parallel to the horizontal projection of the axis, reducing the case to that of Art. 166. The rotation having been effected on the new vertical plane, the measurements may be readily transferred to the primitive planes of projection.

II. ROTATION OF THE LINE.

169. The right line may assume one of three distinct **positions to the axis of rotation:**

(1) It may be parallel to it, describing in its revolution the cylindrical surface.

(2) It may incline to it—the two lying in a common plane—describing in its revolution the cone of one or two nappes.

(3) It may have any position not in the plane of the axis, describing in its revolution the hyperboloid of one nappe.

If the revolving line be a curve, it will generate a double-curved surface.

170. PROBLEM.—*To revolve a right line around a vertical axis and through a given angle.*

Let $(a'b', a''b'')$ be the given line, $(x'y', x''y'')$ the axis, and δ the angle (Fig. 160).

During the rotation of the line each point thereof describes a circle parallel to H, the relative positions of line and axis remaining unaltered; hence the horizontal projection in the revolved position preserves the same distance from the foot x'' of the axis. Draw then from x'' a perpendicular $x''o''$ to $a''b''$, thus marking the shortest distance of the given line, and determine the vertical projection o'. With x'' as a centre and $x''o''$ as a radius describe an arc whose measure is the given angle δ, and draw the new horizontal projection $o_1''b_1''$ perpendicular to $x''o_1''$, the second point b_1'' being obtained by measuring from o_1'' a distance equal to $o''b''$.

As each point during the entire rotation preserves a fixed distance from H, parallels to GL drawn through o' and b' determine o_1' and b_1' through which the vertical projection of the revolved line passes.

In Fig. 161 the axis has been assumed to be perpendicular to V.

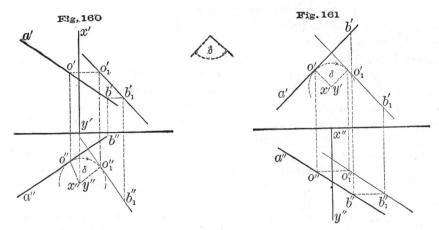

Fig. 160 Fig. 161

171. PROBLEM.—*To revolve a right line until it becomes parallel to either plane of projection.*

Let $(a'b', a''b'')$ be the given line, and $(x'y', x''y'')$ the axis, and let it be required to bring the line parallel to V (Fig. 162). The horizontal projection of the line will be parallel to GL; hence but one point will be required to determine it.

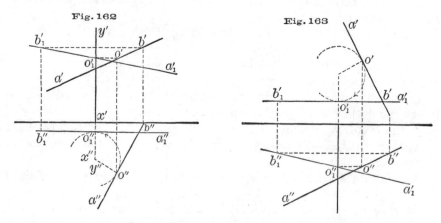

Fig. 162 Fig. 163

In Fig. 163 the line has been brought parallel to H, the axis being perpendicular to V.

172. PROBLEM.—*To revolve a line parallel to the ground-line.*

Combine the two solutions of Art. 171.

173. PROBLEM.—*To revolve a line until it becomes perpendicular to the vertical plane.*

Revolve the line until parallel to H, then turn it about an axis perpendicular to H until its horizontal projection assumes a position perpendicular to GL.

174. PROBLEM.—*To revolve a line until it becomes perpendicular to the horizontal plane.*

Revolve the line until parallel to *V*, then turn it about an axis perpendicular to *V* until its vertical projection assumes a position perpendicular to *GL*.

In the two preceding cases the construction will be simplified by passing the axis through a point of the line.

III. ROTATION OF THE PLANE.

175. A plane is revolved when two of its lines, a line and a point or three points are turned about an axis.

176. When the axis is a line of the revolving plane, the determination of a single point in the new position will be sufficient, inasmuch as, the axis always remaining a line of the plane whatever the position assumed, a point and a line afford the requisite data for that plane.

177. When the axis is parallel to the revolving plane it is to be observed that—

(1) But one line in the plane is both parallel to the axis and measures the shortest distance between the two.

(2) The line so determined describes, during the rotation, a cylinder to which the plane is constantly tangent.

(3) The trace of the plane is tangent to the base of the cylinder.

178. When the axis intersects the revolving plane, the determination of a single line in the new position will fix the position of the plane, the point of intersection between the axis and the plane in connection with that line affording the necessary data.

In this position it is to be observed that—

(1) One line of the given plane is cut by the plane which passes through the axis and is perpendicular to that plane.

(2) This line describes, during rotation, a cone whose vertex is the point of intersection of the axis with the given plane.

(3) The revolving plane is constantly tangent to this cone, and its trace is tangent to the base.

179. PROBLEM.—*To revolve a plane around a vertical axis through a given angle.*

Let $(a'b', a''b'')$ and $(c'd', c''d'')$ be two horizontals of the given plane (Fig. 164), $(x'y', x''y'')$ the axis, and δ the angle.

The relative positions of the horizontals to each other and to the axis are not altered during the rotation, each point of the lines describing a circle parallel to *H*. Draw, then, $x''c''$ and turn it through the given angle, bringing a'' to a_1'', and c'' to c_1'', and determining the horizontal projections of the revolved lines.

The vertical projections coincide with those of the original position.

Should the traces of the plane be given (Fig. 165), two horizontals may again be taken, one of which, by preference, is the horizontal trace *HM*, which will remain during the entire rotation a line of the horizontal plane. Letting fall from x'' a perpendicular to *HM*, the point of intersection p'' describes, during the rotation, a circle to which the trace is constantly tangent. Measure the angle δ and draw $HM_{,}$, the new position of the horizontal trace, thus determining the ground-point o of the vertical trace. A second point may be obtained by revolving any

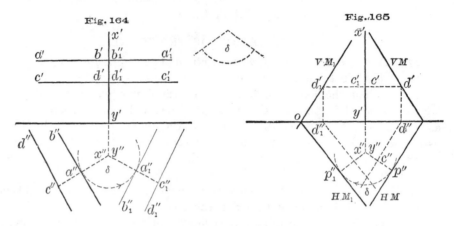

horizontal ($c''d''$, $c'd'$) of the plane and finding its vertical piercing-point (d_1', d_1'').

180. PROBLEM.—*To revolve a plane around an axis perpendicular to the vertical plane.*

In Figs. 166 and 167 the constructions are precisely similar to those

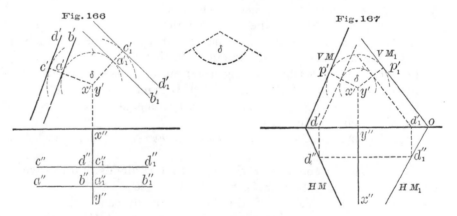

of the two preceding cases, but reversed on the respective planes of projection in accordance with the changed position of the axis.

IV. RABATTEMENT.

181. As it is desirable to avoid constructions which, by the position of the object in space, may appear confused, and as it is also frequently necessary in practice to determine the true size, shape and position of such an object, or of its separate parts, this end is attained by the method of rotation termed *rabattement*.

By way of illustration, assume the revolving object to be a plane;

then, should this plane be revolved around any line which is parallel to either coördinate plane until it is parallel also to that plane, or should it be revolved around any line lying in either coördinate plane until it coincides with that plane, the rotation is by *rabattement*.

By *counter-rotation* is understood the restoration of the plane to its original position.

182. PROBLEM.—*Rabattement of a plane around a vertical, the plane being perpendicular to the vertical plane.*

Let $(x'y', x''y'')$ be the vertical and (a', a'') a point of the plane (Fig. 168) in its original position.

The radius of rotation drawn as a perpendicular from the point (a', a'') to the vertical or axis is parallel to H, and therefore projected in $a''c''$ in its true size. After rotation $a''c''$ becomes parallel to V, is projected on it in its true size, and on H in a line parallel to GL. Hence, making $c'a_1'$ equal to $c''a''$ and drawing the ordinate, (a_1', a_1'') are the projections of the point by *rabattement*.

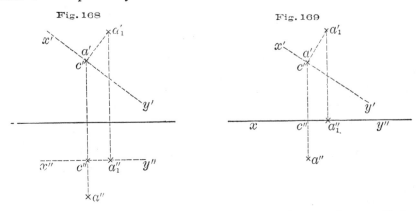

Fig. 168   Fig. 169

The point so found and the fixed vertical are data sufficient to determine the plane in its new position. Should the vertical coincide with the vertical plane (Fig. 169), the *rabattement* will be around the vertical trace of the plane as an axis.

Conversely, if the *rabattement* has been effected, the original position of the object may be ascertained by reversing the operations.

Thus, having the point (a_1', a_1''), by letting fall from this point (Fig. 168) a perpendicular to the vertical $(x'y', x''y'')$, by restoring the plane of the radius and axis to its original position perpendicular to V, and by laying off the distance $c''a''$ equal to $c'a_1'$, (a', a'') are the projections of the point after the counter-rotation has been made.

183. PROBLEM.—*Rabattement of a plane around one of its horizontals.*

Let $(x'y', x''y'')$ be the horizontal and (a', a'') a point of the plane in its original position (Fig. 170).

The radius of rotation drawn as a perpendicular from the point (a', a'') to the horizontal or axis is parallel to V, and therefore projected in $a'c'$ in its true size. After rotation $a''c''$ becomes parallel to H, is projected on it in its true size, and on V in a line parallel to GL. Hence, making $c''a_1''$ equal to $c'a'$ and drawing the ordinate, (a_1'', a_1') are the projections of the point by *rabattement*. The point so found and the fixed horizontal are data sufficient to determine the plane in its new position.

Should the horizontal coincide with H (Fig. 171), the rabattement will be around the horizontal trace of the plane as an axis.

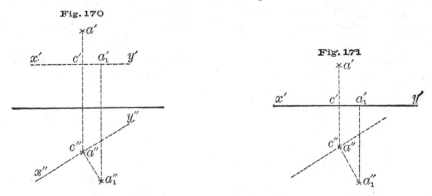

Conversely, if the *rabattement* has been effected, the original portion of the object may be ascertained by reversing the operations.

184. When the plane of the object inclines to both planes of projection, the construction may be effected either by bringing the plane parallel to one of the planes of projection or by a change of ground-line.

The following examples will serve to illustrate the variations in solution in connection with the method of counter-rotation.

185. PROBLEM.—*Through a given point of a plane to draw a rectangle in that plane.*

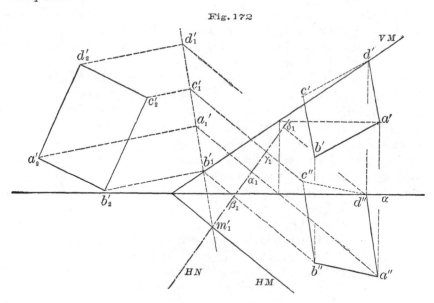

Fig. 172

Let (VM, HM) be the plane, and a' the vertical projection of the point (Fig. 172).

Determine the horizontal projection a'' of the point by means of a horizontal, and assume a new vertical plane perpendicular to the given plane; HN is its horizontal trace or ground-line perpendicular to HM. Since the planes M and N are at right angles, the new vertical projection a_1', found by making $\alpha_1 a_1'$ equal to $\alpha a'$, is a point in the line of intersection or trace between them; the second point being m_1', where the horizontal traces cross each other.

Revolve the plane M around the line $m_1' a_1'$ as an axis into the new vertical plane; find a_2' by making $a_2' a_1'$ equal to $\alpha_1 a''$, and construct the required rectangle $a_2' b_2' c_2' d_2'$.

By a counter-rotation bringing the plane of the rectangle back to its original position, $b_1' d_1'$ is the new vertical projection, from which the primitive horizontal may be found by making $\beta_1 b''$ equal to $b_1' b_2'$, $\gamma_1 c''$ equal to $c_1' c_2'$, and $\delta_1 d''$ equal to $d_1' d_2'$.

186. PROBLEM.—*To construct a triangle on any line of a given plane.*

Let (VM, HM) be the given plane (Fig. 173), and $a'b'$ the vertical projection of the line.

Determine the horizontal projection $(a''b'')$ by means of horizontals, and assume an auxiliary plane S, perpendicular to V and to the given plane M; VS is its vertical trace or ground-line. The intersection of S

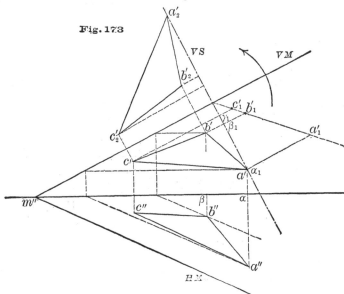

Fig. 173

and M passes through the supplementary projection $(a_1' b_1')$ found by making $\alpha_1 a_1'$ equal to $\alpha a''$, and $\beta_1 b_1'$ equal to $\beta b''$.

By *rabattement* around the vertical trace VM, the points a and b describe circles parallel to S, the projections of which are perpendicular to VM; hence $a_2' b_2'$ is the revolved projection of the line upon which the triangle may be constructed.

By counter-rotation c_2' falls in supplementary projection at c_1', from which the primitive projections (c', c'') are readily obtained.

CHAPTER VIII.

DISTANCES AND PERPENDICULARS.

I. DISTANCE OF TWO POINTS.

188. PROBLEM.—*To determine the distance between two points given by their projections.*

The distance between two points is measured by the right line which joins them.

(1) Let (a', a'') and (b', b'') be the given points (Fig. 174). Bring the plane, horizontally projecting the line joining the two points, parallel to V by *rabattement* around any vertical, preferably that passing through the point (a', a''). This point remains fixed during the rotation, and (b', b'') assumes the position (b_1', b_1''); $a'b_1'$ is therefore the distance required.

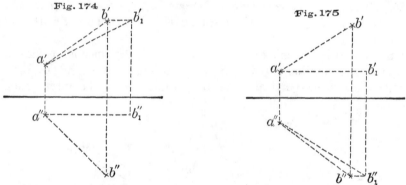

(2) Bring the plane, vertically projecting the line which joins the two points, parallel to H by *rabattement* around any horizontal, preferably that passing through point (a', a'') (Fig. 175). This point remains fixed during the rotation, and (b', b'') assumes the position (b_1', b_1''); $a''b_1''$ is therefore the distance required.

(3) Bring the plane (Fig. 176), horizontally projecting the line which

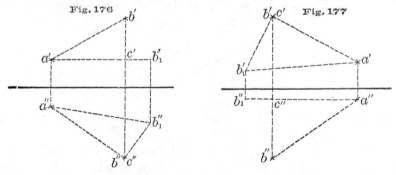

joins the points, parallel to H by *rabattement* around any horizontal, preferably that passing through (a', a''). This point remains fixed, and the point (b', b'') assumes the position (b_1', b_1'') by making the perpendicular $b''b_1''$ equal to $c'b'$; hence $a''b_1''$ is the distance required.

(4) Bring the plane (Fig. 177), vertically projecting the line which joins the two points, parallel to V by *rabattement* around any vertical, preferably that passing through (a', a''). This point remains fixed, and the point (b', b'') assumes the position (b_1', b_1'') by making the perpendicular $b'b_1'$ equal to $c''b''$; hence $a'b_1'$ is the distance required.

189. PROBLEM.—*Upon a given line to measure a given distance from either extremity.*

Let (a', a'') be the extremity from which the measurement is to be made (Figs. 178, 179), and (b', b'') any other point of the given line.

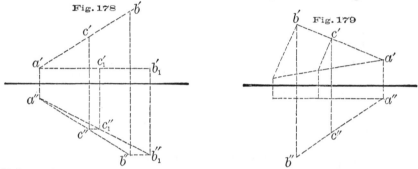

Bring the line by any of the preceding four methods parallel to either coördinate plane, and measure upon the projection so determined the required length. By a counter-rotation restore the dividing point (c_1', c_1'') to the primitive projections; $(a'c', a''c'')$ is the distance sought.

II. DISTANCE OF POINT FROM LINE.

190. PROBLEM.—*To determine the perpendicular between a point and a line given by their projections.*

The point and line fixing the position of a plane, their distance from each other may be found by the *rabattement* of that plane.

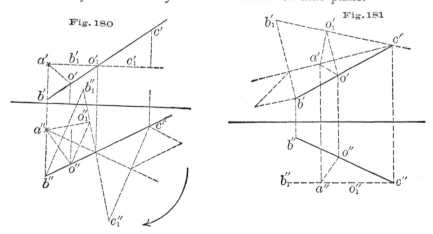

(1) Let (a', a'') be the given point, and $(b'c', b''c'')$ the given line (Fig. 180)

Bring the plane of these two by *rabattement* around a horizontal, preferably that which passes through the point (a', a''). During rotation this point remains fixed, and the line bc assumes the position $(b_1''c_1'', b_1'c_1')$ (Art. 183); hence, letting fall a perpendicular $(a''o_1'')$ upon $b_1''c_1''$, $a''o_1''$ is the horizontal projection of the perpendicular sought.

By counter-rotation the foot of the perpendicular (o_1', o_1'') falls in (o', o''), and hence $(a'o', a''o'')$ are the primitive projections of the line let fall from (a', a'') upon the given line.

In Fig. 181 the *rabattement* has been effected around a vertical $(a'c', a''c'')$ passing through the given point (a', a'').

During rotation this point remains fixed, as does likewise (c', c'') of the given line; hence $b_1'c'$ will be the line revolved, and $a'o_1'$ the perpendicular sought.

By counter-rotation $(a'o', a''o'')$ are the projections of the perpendicular let fall from the given point upon the line.

2d Solution.—The problem may also be solved by passing through the given point a plane perpendicular to the given line, when the line joining this point with the point of intersection between the given line and auxiliary plane determines the required perpendicular.

Thus in Fig. 182 let (a', a'') be the given point, and $(b'c', b''c'')$ the given line.

To fix the position of the auxiliary plane three steps are necessary:

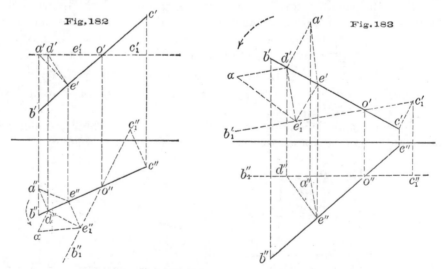

to pass any plane through the given line, to let fall upon this plane a perpendicular from the given point, and to draw from the foot of the perpendicular a second perpendicular to the given line. The two perpendiculars determine the required plane.

Hence, assume the plane horizontally projecting the line in $b''c''$ as the plane of the line; the perpendicular let fall upon it from a in space gives the projections $(a''d'', a'd')$. From the foot (d', d'') of this perpendicular draw a perpendicular to $(b'c', b''c'')$, which may be effected by the *rabattement* of the projecting plane of the line around one of its horizontals, preferably that which passes through the point (d', d''). During rotation the horizontal remains fixed; and the line bc assumes the position $(b_1''c_1'', b_1'c_1')$; hence $(d''e_1'', d'e_1')$ is the second perpendicular sought.

By counter-rotation (e_1'', e_1') falls at (e'', e'); then $(a''d'', a'd')$ and $(d''e'', d'e')$ are the two lines which determine the position of the required plane, and $(a'e', a''e'')$ is the perpendicular drawn from the given point to the given line.

The distance of the point (a', a'') from the given line is equal to the hypothenuse of a right-angled triangle, of which the base is the distance $a''d''$ of the point from the plane of the line, and the perpendicular the distance $d''e_1''$ of its foot from the given line; hence drawing $d''a$ perpendicular to $d''e_1''$ and equal to $d''a''$, ae_1'' is the distance of the point from the line.

In Fig. 183 the plane projecting the line on V in $b'c'$ being assumed as the plane of the line, the operations are a repetition of those of the preceding case.

191. PROBLEM.—*From a given point as a vertex, to construct a triangle on a given line.*

Let (a', a'') be the given point (Fig. 184), and $(b'c', b''c'')$ the line.

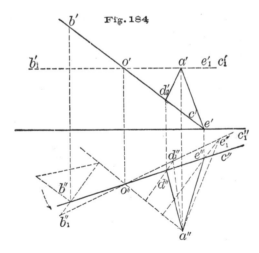

Fig. 184

Bring the plane of these two by *rabattement* around a horizontal $(a'o', a''o'')$, by the operations of the preceding problems. From a'' as a centre and with radii equal to the length of the given sides of the triangle, describe arcs intersecting $b_1''c_1''$ in d_1'' and e_1''; $a''d_1''e_1''$ is the triangle in its true size.

By counter-rotation the points of the triangle are projected at (e'', e') and (d'', d'); hence $(a''d''e'', a'd'c')$ are the projections of the required triangle on the primitive planes.

In Fig. 185 the *rabattement* has been effected around a vertical $(a'o',$ $a''o'')$; hence $b_1'c_1'$ is the revolved line to which the measurements from

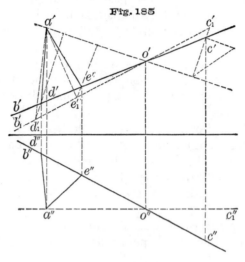

Fig. 185

the given point (a', a'') are to be made. The triangle $a'd_1'e_1'$ is represented in its true size, and $(a'd'e', a''d''e'')$ are its projections on the primitive planes after the counter-rotation.

III. DISTANCE OF POINT FROM PLANE.

192. The distance of a point from a plane is measured by the perpendicular let fall from the point upon that plane. This perpendicular may be found by passing through the point any auxiliary plane perpendicular to the given plane, by determining the line of intersection between the two planes, and by drawing to this line a perpendicular from the point. The case resolves itself, therefore, into that of Art. 190.

193. PROBLEM.—*To determine the distance between a point and a plane.*

The following are various solutions showing the application of preceding principles.

(1) Let (a', a'') be the given point, $(b'c', b''c'')$ a horizontal, and (d', d'') a point of the given plane (Fig. 186).

Through the point (a', a'') pass a plane perpendicular to the horizontal $(b'c', b''c'')$; its horizontal trace by Art. 104 passes through a'' and is perpendicular to $b''c''$ at c'', which is a point of the line of intersection between this auxiliary plane and the given plane. The horizontal $(d'e', d''e'')$ determines the second point (e', e''), whence $(c'e', c''e'')$ is the line of intersection sought.

By the *rabattement* of the auxiliary plane—which contains the given point and the line of intersection—around a horizontal, preferably that which passes through the point (c', c''), the points (e', e'') and (a', a'') fall at (e_1'', e_1') and (a_1'', a_1') respectively; hence $a_1''o_1''$ is the *rabattement* of the perpendicular drawn from the point (a', a'') to the given plane.

By counter-rotation o_1'' is projected in o'', and $(a''o'', a'o')$ are the primitive projections of the required distance.

(2) Let (VM, HM) be the given plane, and (a', a'') the point (Fig. 187).

Let fall from the point a perpendicular upon the plane (Art. 103), and find the piercing-point thereon; the distance between the two

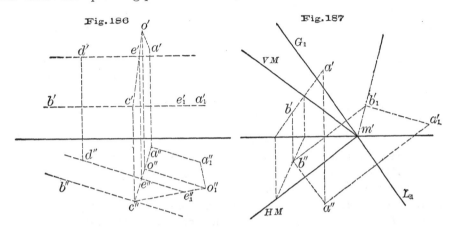

Fig.186

Fig.187

points measures the distance from the plane. To determine this, assume a new vertical plane, of which G_1L_1 is the ground-line perpendicular to the given plane, and hence parallel to the line $(a'b', a''b'')$; $a_1'b_1'$ is the new vertical projection marking the real length of the line, and $m'b_1'$ the new vertical trace to which it will be perpendicular.

(3) Let (VM, HM) be the given plane, and (a', a'') the point (Fig. 188).

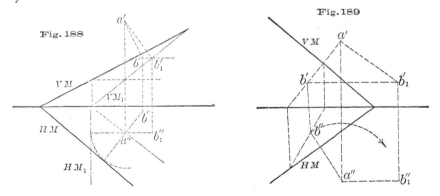

Fig. 188

Fig.189

Bring the plane M perpendicular to V by rotation around a vertical axis passing through the given point (a', a''); (HM_1, VM_1) are the traces in the new position (Art. 179). During the rotation the point (a', a'') remains fixed, and the line $(a'b', a''b'')$ drawn perpendicular to M becomes parallel to V; hence $a'b_1'$ drawn perpendicular to VM_1 is the required perpendicular in its true length.

Restoring the plane to its original position, $(a'b', a''b'')$ are its pro-jections in the original position.

(4) Let (VM, HM) be the given plane, and (a', a'') the point (Fig. 189).

From (a', a'') let fall a perpendicular upon the plane, find its piercing-point thereon, and revolve the line $(a'b', a''b'')$ parallel to either coör-dinate plane (Art. 171).

194. PROBLEM.—*Through a given point in a line to pass a plane perpendicular to that line.*

The given point is a point of the required plane; hence a horizontal or vertical passed through it determines by its piercing-point the required traces.

195. PROBLEM.—*Through a given point to pass a plane perpendicular to a given plane.*

Through the point let fall a perpendicular upon the plane. Every plane passing through this line will be in the required position.

196. PROBLEM.—*To determine the distance between two parallel planes.*

Resolved into the problem Art. 193.

197. PROBLEM.—*To erect a right prism standing upon a given plane.*

Let (HM, VM) be the given plane, and $a'b'c'$ the vertical projection of the base of the prism (Fig. 190). Find the horizontal projection $a''b''c''$ by means of horizontals, and draw indefinite lines perpendicular to the

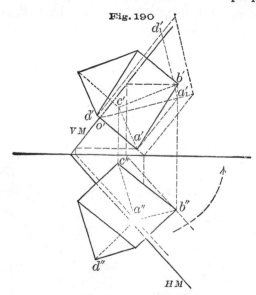

Fig. 190

plane for the projections of the edges in position. Rotate any plane, projecting an edge on V, by *rabattement* around its vertical trace, as, for example, the projecting plane of the edge (ad) in space; the point a_1' is the *rabattement* of a of the base, and $a_1'o'$ that of the line of in-tersection between the plane M and the projecting plane. But the edge is perpendicular to every line of the given plane passing through its foot; hence $a_1'd_1'$ drawn perpendicular to $o'a_1'$ and made equal to the required altitude of the prism is the *rabattement* of the edge ad.

By counter-rotation d_1' falls at d'; hence $(a'd', a''d'')$ are the projections of the edge on the primitive planes. Lay off from (b', c') and (b'', c''), respectively, distances equal to $a'd'$ and $a''d''$, and complete the projections of the prism by joining the extremities of the edges so found.

198. PROBLEM.—*Given a plane and the rabattement of the base of a right pyramid standing on that plane, to determine the projections of the pyramid after counter-rotation.*

Let (VM, HM) be the given plane (Fig. 191), and $a_2''b_2''c_2''d_2''$ the *rabattement* of the base around the horizontal trace HM.

The ordinates $\alpha a_2''\beta b_2''$, etc., measure the perpendicular distances of the points of the base from the trace HM both before and after rotation; hence, if the plane be counter-rotated, these distances will be the hypothenuses of right-angled triangles, of which the perpendiculars are the heights of the points above H, and the bases the distances of their horizontal projections from the trace HM.

Assume a new vertical plane which shall **pass through** $\epsilon e_2''$ **the middle**

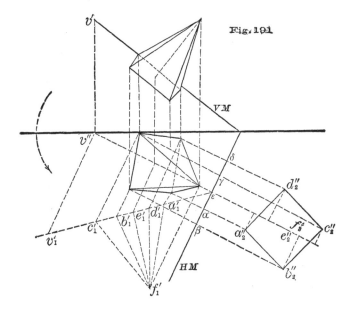

Fig. 191

point of the base and be perpendicular to the plane M; its horizontal trace $e_2''v''$ is perpendicular to HM. By *rabattement* around its horizontal trace, the line of intersection with the given plane falls at $\epsilon v_1'$ and contains the new vertical projection $a_1'b_1'c_1'd_1'$ of the base. To determine this projection lay off upon the line $\epsilon v_1'$ the distances of the points a_2'', b_2'', etc., from HM. At e_1' erect the given altitude $e_1'f_1'$, and complete the new vertical projection of the pyramid, from which the primitive projections may readily be found by the usual methods.

IV. SHORTEST DISTANCE BETWEEN TWO LINES.

199. A line may be a *common perpendicular* to two right lines either in *direction* or *position*.

It is a *perpendicular in direction* when, should both lines be moved parallel to their original positions respectively, and towards the assumed line, they will intersect it at *right angles*.

Thus, if any plane (*VM, HM*) be assumed as perpendicular to the first line (*a'b', a''b''*) (Fig. 192), then any line parallel to that plane will be a perpendicular in direction to the given line; in like manner, if any plane (*VP, HP*) be assumed as perpendicular to the second line, then any line parallel to that plane will be a perpendicular in direction to the given line. Hence the direction sought is parallel to both planes, or to their line of intersection (*m'p', m''p''*).

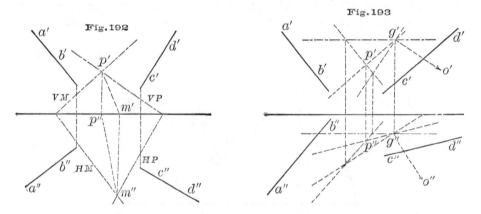

The perpendicular in direction may also be fixed in its position by determining a plane parallel to the two given lines, when any perpendicular to the plane so found is the direction sought.

Thus, in Fig. 193, through any assumed point (*p', p''*) lead lines parallel to the given lines, and find a vertical and a horizontal of the plane which they determine (Art. 82); the perpendicular (*g'o', g''o''*) to this plane is the direction sought.

200. The *common perpendicular in position* to two right lines is that line which *intersects both* at right angles.

When the two lines are parallel they lie in the same plane; hence, any line which intersects them must also be a line of that plane. In this special case any number of lines may be perpendicular in position.

When the lines are not parallel there is but one perpendicular in position. As the perpendicular in direction is parallel to the perpendicular in position, planes passing through the two lines and parallel, respectively, to the perpendicular in direction must intersect in the perpendicular in position. Thus, in Fig. 194, the perpendicular in direction *p'm',·p''m''* having been determined as in Fig. 192, through each of the lines *ab* and *cd* in space pass a plane parallel to this line (Art. 94). The line of intersection between the two planes being parallel to $(p'm', p''m'')$, a single point l'', determined by means of the horizontals

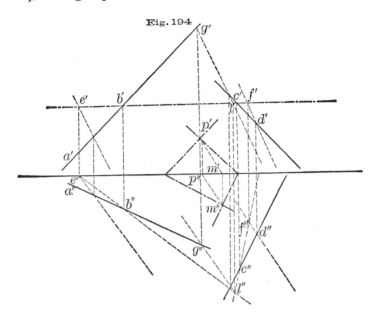

Fig. 194

eb and *cf*, is sufficient to determine $(g'l', g''l'')$, the position of the common perpendicular sought. The perpendicular in position having been found, the points in which it intersects the given lines are the extremities of the line which measures the shortest distance between them.

201. PROBLEM.—*To determine the line measuring the shortest distance between two right lines not in the same plane.*

Let *ab* and *cd* be the given lines (Fig. 195).

(1) Pass through the line *ab* a plane parallel to *cd*; *HM* and *VM* are its traces.

(2) From any point of *cd*, as *p*, let fall a perpendicular upon the plane *M*, and find its piercing-point (m', m'') thereon.

(3) Through (m', m'') draw a parallel to $(c'd', c''d'')$; then are $(m'a', m''a'')$ the projections of *cd* on the plane *M*; hence, *ma* and *ab* lying in the same plane intersect each other in the point (a', a'').

(4) From the point (a', a'') erect a perpendicular to the plane *M*, which is the perpendicular in position sought. For $(a'd', a''d'')$ is perpendicular to $(a'b', a''b'')$ and $(a'm', a''m'')$, since they pass through its foot (a', a'') and lie in the plane *M*, to which it is perpendicular; hence it is a projecting line of the plane projecting *cd* upon *M*, and must intersect *cd* in the point *d*.

The true length of the line $(a'd', a''d'')$ may be determined by Art. 193.

202. When one of the lines is perpendicular to either coördinate plane the construction is greatly simplified.

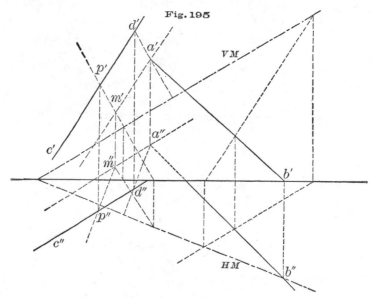

Fig. 195

Thus in Fig. 196 the line *ab* is vertical; hence the plane passed through *cd* parallel to it is also vertical, and the perpendicular in position horizontal. The plane passed through *ab* parallel to the line of shortest distance gives a horizontal trace $a''e''$ perpendicular to $c''d''$; hence ($a'e'$, $a''e''$) are the projections of the line of shortest distance; and since this line is horizontal, $a''e''$ is its true measurement.

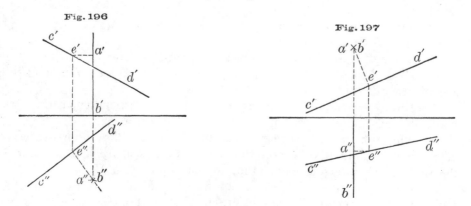

Fig. 196   Fig. 197

In Fig. 197, as the plane passed through *cd* parallel to *ab* is perpendicular to *V*, the perpendicular in position is parallel to *V*.

The plane passed through *ab* parallel to the line of shortest distance gives a vertical trace $a'e'$ perpendicular to $c'd'$; hence ($a'e'$, $a''e''$) are the projections of the shortest distance; and since the line is parallel to *V*, ($a'e'$) is its true measurement.

203. Should both lines be parallel to either coördinate plane the construction will be still further simplified.

Thus, in Fig. 198, the lines *ab* and *cd* being parallel to *H*, the plane passed through *ab* is horizontal, and the perpendicular in direction is vertical; hence the planes passing through the given lines and parallel to this perpendicular are also vertical, and give *e′f′*, the vertical projection of the line of shortest distance, in its true measurement.

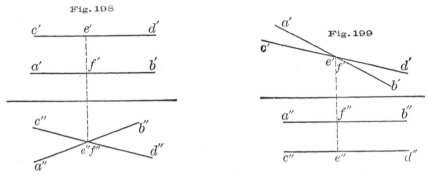

In Fig. 199 similar reasoning will show (*e′f′*, *e″f″*) to be the projections of the line of shortest distance, and *e″f″* its true measurement.

204. When the two lines lie in the same plane they must either intersect or be parallel.

In the first case the line of shortest distance passes through the point of intersection, and hence is reduced to a point.

In the second case the plane passing through one line parallel to the other will be indeterminate and the application of the general rule will fail. But the shortest distance may readily be found by letting fall from any point of either line a perpendicular to the other (Art. 190).

205. PROBLEM.—*Given the horizontal projections of two lines and the two projections of their line of shortest distance, to determine the vertical projections.*

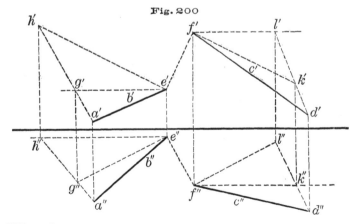

Let *a″b″* and *c″d″* (Fig. 200) be the horizontal projections of the two lines, and (*e′f′*, *e″f″*) the projections of the line of shortest distance.

The line *ab* is perpendicular to *ef*, and hence lies in a plane passing through *e* and perpendicular to *ef*. Determine a horizontal (*e'g'*, *e''g''*) and a vertical (*e'h'*, *e''h''*) of this plane, and assume any third line of this same plane, as, for example, the one whose horizontal projection is *a''h''*; find its vertical projection *h'g'*, and *a'* is projected in it.

By a similar process of reasoning the vertical projection *c'd'* may be determined.

206. PROBLEM.—*Given the projections of a line and the horizontal projections of a second line and the line of shortest distance, to determine their vertical projections.*

Let (*a'b'*, *a''b''*) be the given line (Fig. 201), and *e''f''*, *c''d''* the hori-

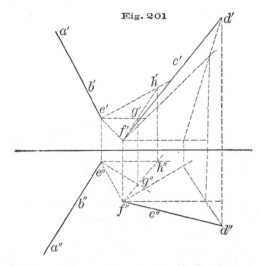

Fig. 201

zontal projections of the shortest distance and the second line respectively.

The line *ef*, being perpendicular to *ab*, lies in a plane passing through *e* and perpendicular to *ab*. Determine a horizontal (*e''g''*, *e'g'*) and a vertical (*e'h'*, *e''h''*) of this plane, assume any third line of the plane, as that whose horizontal projection is *f''h''*, find *h'* and *g'*, and *f'* lies in the line which connects them.

The vertical projection *c'd'* of the second line may be found as in the preceding case.

CHAPTER IX.

ANGLES.

I. ANGLES BETWEEN RIGHT LINES.

207. Right lines are either parallel or intersect, or without being parallel do not intersect. In every case the plane angle obtained by passing through a point lines parallel to the lines in space is termed the *angle of the lines.*

To determine these angles, therefore, it is simply necessary to bring the plane of the angle parallel to either coördinate plane.

208. PROBLEM.—*To determine the angle between two lines whose projections are given.*

The lines may lie in the same plane, which will be indicated when the points of intersection of the projections lie in a common perpendicular to GL (Art. 53).

Let $(a'b', a''b'')$ and $(c'd', c''d'')$ be the given lines (Fig. 202), and (o', o'') their point of intersection.

Find the horizontal piercing-points a'' and d'', through which passes the horizontal trace of the plane of the lines. Turn this plane into H by rotating (o', o'') around $a''d''$ as an axis (Art. 163), when α is the required angle.

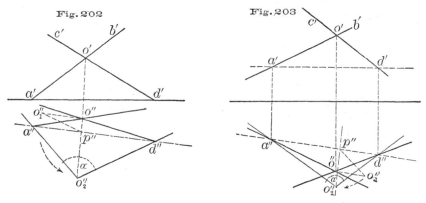

In Fig. 203 the piercing-points are not immediately available. The solution in this case is effected by the *rabattement* of the plane of the angle around any horizontal, as $(a'd', a''d'')$.

Should one of the intersecting lines *ab* be horizontal (Fig. 204), the horizontal trace $c''f''$ of the plane of the angle will be parallel to $a''b''$ (Art. 71). Around this trace as an axis revolve the point (o', o''), and α is the angle sought.

If the point of intersection lies in *H* (Fig. 205), through any point (e', e'') of one line lead a line ($e'g'$, $e''g''$) parallel to the other; the angle between these two, which is equal to the angle sought, may be determined as in Fig. 202.

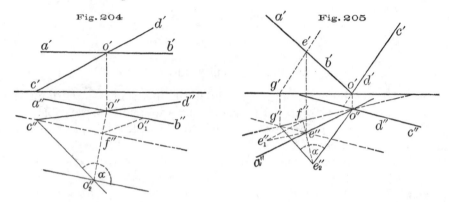

209. PROBLEM.—*To determine the angles which a line in space makes with its projections.*

Bring the plane, projecting the line horizontally, by *rabattement* around $a''b''$ (Fig. 206), and as the line *ab* in space and its projection $a''b''$ lie in a common plane, their prolongations intersect in a point (a', a'') and include the angle α, or the angle of inclination to the horizontal plane of projection.

The *rabattement* measures *the true length of the line in space.*

A similar construction determines the angle β for the vertical plane.

In Fig. 207 the projecting plane of the line is brought by *rabattement* around either a horizontal ($a''c''$, $a'c'$) in order to determine α, or

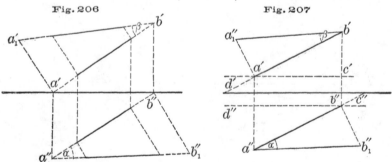

around a vertical ($b'd'$, $b''d''$) to determine β.

210. PROBLEM.—*To determine the angle between the traces of a given plane.*

The sides of the angle or the traces being respectively a vertical and a horizontal of the plane (Fig. 208), the solution is effected as in Art. 208.

211. PROBLEM.—*The projections of two intersecting lines being given, to bisect or otherwise divide their angles.*

Let $(a'b', a''b'')$ and $(c'd', c''d'')$ be the given lines (Fig. 209).

Determine the angle α between them, as in the preceding cases, and draw the lines $o_2''p''$ and $o_2''q''$ which divide the angles in the required ratios. As these lines lie in the plane of the angle, their feet (p'', p')

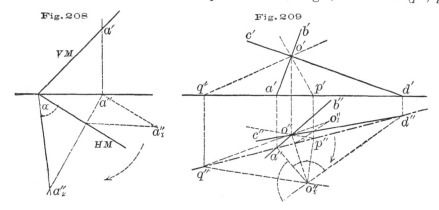

Fig. 208

Fig. 209

and (q'', q') must be contained in the horizontal trace $a''d''$ of that plane; hence, during counter-rotation, they remain fixed in position, and the projections of the bisecting or other dividing lines sought must pass through the points (p', p'') (q', q'') and the vertex (o', o'') of the angle.

212. PROBLEM.—*Through a given point to pass a line which intersects a given line at any angle.*

Let (a', a'') be the given point, and $(b'c', b''c'')$ the given line (Figs. 211, 212).

Bring the plane, which they determine, by *rabattement* around a hori-

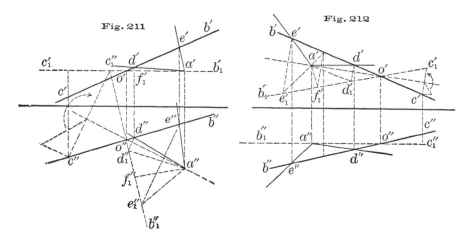

Fig. 211

Fig. 212

zontal $(a'o', a''o'')$ (Fig. 211), or a vertical $(a'o', a''o'')$ (Fig. 212), passing through the point (a', a''). During rotation the point remains fixed, and the line $(b'c', b''c'')$ assumes the position $(b_1''c_1'', b_1'c_1')$. Through (a', a'') lead two lines which make with the perpendicular $(a''f_1'', a'f_1')$, let fall from this point upon the line $(b_1''c_1'', b_1'c_1')$, the complement of the required angle, and counter-rotate the plane.

The points (e_1'', e_1') and (d_1'', d_1') are now projected at (e'', e') and (d'', d'); joining them with (a', a''), the lines $(a''e'', a'e')$ and $(a''d'', a'd')$ are the lines sought.

II. ANGLES BETWEEN LINES AND PLANES.

213. The angle between a line and a plane is measured by *the angle which the line makes with its projection on that plane.*

214. PROBLEM.—*To determine the angle between a line and a plane.*

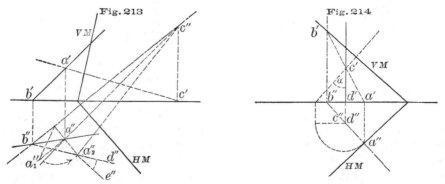

Fig. 213   Fig. 214

Let $(a'b', a''b'')$ be the given line, and (VM, HM) the plane (Fig. 213.)

From any point (a', a'') of the line let fall a perpendicular upon the plane, and find the angle $c''a_2''d''$ between these two lines (Art. 208). The *complement*, $d''a_2''e''$, of this angle is the required angle.

When the given line is a vertical (Fig. 214), pass through the line a plane perpendicular to the horizontal trace HM of the given plane; the two planes intersect in a line $(a'b', a''b'')$, which contains the projection of the given line upon the plane M. The angle between the line and its projection may be determined by rotating $(a'b', a''b'')$ around $(c'd', c''d'')$ as an axis until it is brought parallel to V, when α is the required angle.

When the given line is the ground-line (Fig. 215), let fall from any point x a perpendicular to the given plane (HM, VM), and find its piercing-point (c', c'') thereon; $o'c', o'c''$ are the projections of the ground-line upon the plane M. But $o''x''$ is the hypothenuse of a triangle, right-angled at (c', c''), of which $o''c''$ is the horizontal projection of the base, and $x''c''$ of the perpendicular; hence it may be inscribed in a semi-circle described around $x''o''$ as a diameter. Rotating the semicircle into H, the point c falls at c_1'', and $x''o''c_1''$ is the angle sought.

215. PROBLEM.—*To pass through a given point a line making any angle with a plane.*

Let (a', a'') be the given point, (VM, HM) the plane, and α the angle (Fig. 216).

From (a', a'') let fall a perpendicular upon M, find its piercing-point

(b', b'') thereon, and draw through it any line of the plane, preferably a horizontal ($b'c'$, $b''c''$) or a vertical. The two lines drawn through a, and making the required angle with ($b'c'$, $b''c''$), are the required lines. The problem is then resolved into a special case of Art. 212.

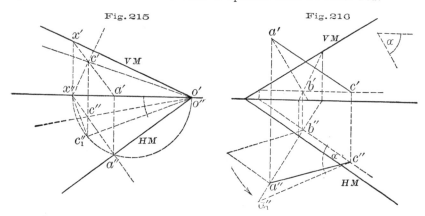

Fig. 215   Fig. 216

216. *Lines of Greatest Declivity.*—Those lines in a plane which measure the greatest possible angle with any second plane are termed the lines of greatest declivity. As they are perpendicular to the line of intersection between the two planes (Art. 71), it follows that those which measure the greatest angle to H are perpendicular to the horizontals, and those which measure the greatest angle to V are perpendicular to the verticals. Hence any plane cutting a given plane perpendicular to its horizontals or verticals will determine respectively the lines of greatest declivity to H or V.

217. PROBLEM.—*To determine a plane when its line of greatest declivity is given.*

Let ($a'b'$, $a''b''$) be the given line measuring the greatest declivity to H (Fig. 217).

Since this line is perpendicular to the horizontals, lead through any

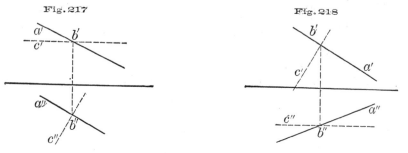

Fig. 217   Fig. 218

point, as (b', b''), of the given line a horizontal of the plane; its horizontal projection $c''b''$ is perpendicular to $a''b''$. By means of these two intersecting lines the plane can readily be determined (Art. 85).

In a like manner when the measure is to V (Fig. 218), the vertical projection of the given line is perpendicular to the verticals.

218. PROBLEM.—*To determine the projections of a line when its inclina-tions to the coördinate planes are given* (Fig. 219).

Let α and β be the given angles, marking respectively the inclinations to H and V.

Take any point (a', a'') and assume that the required line passes through it, thus forming, with its horizontal projection, a triangle which is right-angled at a''. By *rabattement* around $a'a''$ it is shown in its true size at $(a'a''b_1'')$. But $a'b_1''$ is likewise the hypotenuse of the right-angled triangle which the line in space forms with its vertical projection; hence,

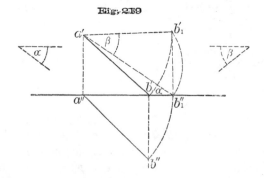

Fig. 219

constructing this triangle, $a'b_1'$ is the length of the vertical projection. With a' as a centre, and with a radius equal to $a'b_1'$, describe an arc cutting GL at b'; $a'b'$ is the vertical projection of the line sought. With $a''b$ as a radius, and with a'' as a centre, describe a second arc intersecting the ordinate let fall from b' in b''; $a''b''$ is the horizontal projection of the line.

III. ANGLES BETWEEN PLANES.

219. The angle between two planes may be reduced to the case of the angle between two lines; for, if from any point assumed within the angle perpendiculars be drawn to the two planes, the angle measured by these perpendiculars will be the *supplement* of the angle of the planes.

The angle is likewise directly measured by the two lines which are cut by an auxiliary plane led perpendicular to the line of intersection between the two planes.

220. PROBLEM.—*To determine the angle between two planes.*

Let (HM, VM) and (HL, VL) be the traces of the given plane (Fig. 220).

Find the projection $(a'b', a''b'')$ of the line of intersection, and pass a plane perpendicular to this line, giving the horizontal trace $d''c''e''$ perpendicular to $a''b''$. This auxiliary plane cuts the given planes in lines which form with $d''c''e''$ a triangle whose vertex lies in the line $(a'b', a''b'')$; but, as this vertex is common to the auxiliary plane and to the plane projecting ab in $a''b''$, it is a point of their line of intersection which passes through c'' and is perpendicular to $(a'b', a''b'')$. Hence, by the *rabattement* of *ab* around *a''b''*, the distance *c''f₁''*, found by drawing a perpendicular from *c''* to *a''b₁''*, determines the revolved projection of the vertex *f₁''* of the triangle. Laying off this measurement from *c''* upon *c''a''*, and drawing *f₂''d''* and *f₂''e''*,. the included angle *α* is the angle sought.

In Fig. 221 the horizontal traces coincide; hence the common trace is the line of intersection between the two planes, and the auxiliary

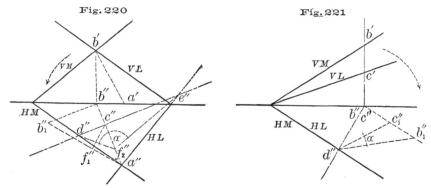

Fig. 220   Fig. 221

cutting plane passed perpendicular to it is likewise perpendicular to *H*. By *rabattement* around *d''b''* the lines of intersection *d''c₁''* and *d''b₁''* are determined, and with them the included angle *α*.

In Fig. 222 the horizontal traces are parallel; hence their line of intersection (*a'b'*, *a''b''*) is parallel to *H*, and the auxiliary cutting plane is perpendicular to it. By *rabattement* around *d''e''* the triangle and the required angle *α* are determined.

In Fig. 223 the traces of the respective planes form continuous lines, and their line of intersection (*a'b'*, *a''b''*) lies in a plane perpendicular to

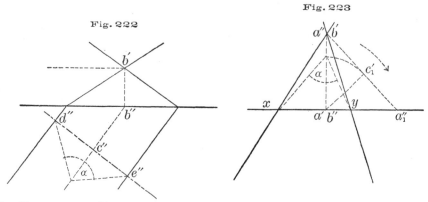

Fig. 222   Fig. 223

GL. Pass an auxiliary cutting plane through the ground-line and perpendicular to (*a'b'*, *a''b''*), intersecting the given planes in the sides of a triangle, the base of which is *xy*, and the line *ab* in a point which is the vertex thereof. Rotate *ab* into *V* around *b'b''* as an axis, and from *b''* let fall the perpendicular *b''c₁'* upon it, thus determining *c₁'*, the revolved projection of the vertex. Lay off from b'' the distance $b''c_1'$ upon $b''a''$, and α is the angle sought.

221. PROBLEM.—*To determine the angles which a given plane makes with the coördinate planes.*

Let (VM, HM) be the given plane (Fig. 224).

The traces being the lines of intersection between the given plane and the coördinate planes, the cutting planes (HL, VL) and (VN, HN) respectively determine the inclinations to H and V. By the *rabattement* of the triangles around the traces HL and VN the required angles α and β are determined.

In Fig. 225 the *rabattement* has been effected around VL and HN respectively.

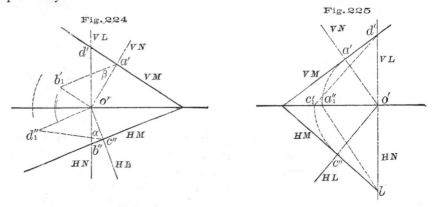

Fig. 224   Fig. 225

222. PROBLEM.—*To determine the planes which contain the line of inter-section between two planes and which bisect their angles.*

Let the planes be given as in Fig. 226.

Find the angle between the two planes (Art. 220), and bisect the angles by means of the projections $(h_1''g'')$ and $(h_1''l'')$ Restoring the triangle $g''h_1''e''$ to its original position, the bisecting lines become lines of the bisecting planes. During this counter-rotation the piercing-points (g'', g') and (l'', l') remain fixed and serve, in connection with the piercing-points (b'', b') and (a'', a') of the line of intersection, to determine the bisecting planes.

223. PROBLEM.—*To find the traces of a plane when its inclination to the coördinate planes is given.*

As the lines of greatest declivity measure the inclinations to the planes, assume that the auxiliary planes cutting these lines pass through a common point a'' in GL (Fig. 227). Again, the auxiliary planes, being perpendicular to the required plane, intersect in a line $a''c$, which is also perpendicular to that plane and to the lines of greatest declivity ($b'c$ and $e''c$) passing through its foot c.

Assume, then, the point (a', a'') as a point of the required plane (Fig. 228); and by *rabattement* around $a'a''$ determine the triangle $a'a''e_1'$, whose hypothenuse $a'e_1'$ measures the greatest declivity α to H. Draw

$a''c_1$ perpendicular to $a'e_1'$, and from a'' as a centre, and with a radius equal to $a''e_1'$, describe an indefinite arc; it is evident that the horizontal trace of the plane will be tangent to it.

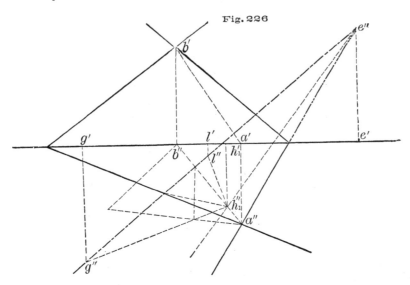

Fig. 226

The line $a''c$ in space (Fig. 227) makes with the trace ($g'g''$) an angle equal to β; hence, if the angle $c_1'a''d_1'$ (Fig. 228) be made equal to β, $a''d_1'$ will equal $g'g''$; and by describing an arc from a'' with $a''d_1'$

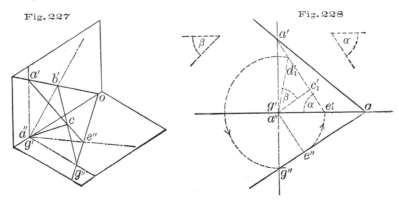

Fig. 227   Fig. 228

as a radius, the point g'' will fall in the ordinate drawn from g'. The horizontal trace passes through g'' and is tangent to the arc $e''e_1'$ at c'', and the vertical trace passes through the assumed point a'.

CHAPTER X.

CHANGE OF POSITION BY COMBINED MOTIONS.

224. In applying the methods of rotation to other than simple geo-metrical surfaces, the graphical constructions will be simplified by the employment of an additional motion from place to place, whereby, instead of having the various projections which result from the change in position overlapping and difficult to disentangle, they are separated, and consequently more readily seen and apprehended. The motion by which this change is effected is termed *a motion of translation or trans-position.*

225. PROBLEM.—*Given a triangle, required to find its projections when the axis is made to incline to both planes of projection.*

(1) Project the triangle in a simple position; for example, with its axis perpendicular to H (Fig. 229, 1).

(2) Let the triangle so placed be moved from its first position towards the right, each point describing a parallel to GL. When far enough removed to avoid interference with the previous position, rotate it around $c_1'c_1''$ until the axis assumes the required position to H. It is evident that, as the object has not changed its position to V, its ver-

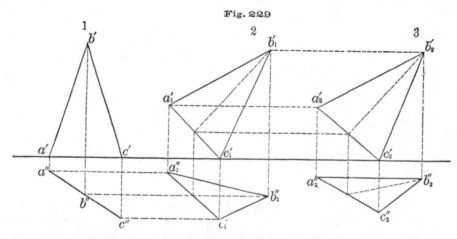

Fig. 229

tical projection in the second position must be an exact counterpart of the vertical projection in the first, the sole difference being in the in-clination to GL.

Hence, measure from c_1' (Fig. 229, 2) a distance $c_1'a_1'$ equal to $c'a'$, making with GL the complement of the given inclination of the axis to H, and complete the second vertical projection $a_1'b_1'c_1'$. As each point, during the combined motion of translation and rotation, has preserved its original distance from V, the horizontal projection of any point will be found at the intersection of the ordinates drawn from the second vertical and first horizontal projections. Thus, a_1'' falls at the intersection of $a''a_1''$ and $a_1'a_1''$, b_1'' at the intersection of $b''b_1''$ and $b_1'b_1''$, etc.

(3) Move the triangle towards the right and rotate it around a vertical line passing through c_2'' until the axis has assumed the required position to V; the horizontal projection $a_2''b_2''c_2''$ will be precisely the same as that in the second position, the difference being in the inclination to GL (Fig. 229, 3). During the combined motions of translation and rotation each point has preserved its distance from H unaltered; hence the vertical projections of the points will be found at the intersections of the ordinates drawn from the horizontal projection of the third position and the vertical of the second. Thus, a_2' falls at the intersection of $a_2''a_2'$ and $a_1'a_2'$, b_2' at the intersection of $b_2''b_2'$ and $b_1'b_2'$, etc.

226. PROBLEM.—*To project a cube in a doubly oblique position.*

(1) Find the projection in a simple position (Fig. 230, 1).

(2) By a motion of translation bring the cube so placed to the right of the first position and rotate it around $c_1''c_1'$ until it assumes the required position to H. During these combined motions the object has not changed its position to V; hence the vertical projection of the second position (Fig. 230, 2) will be precisely the same as that of the

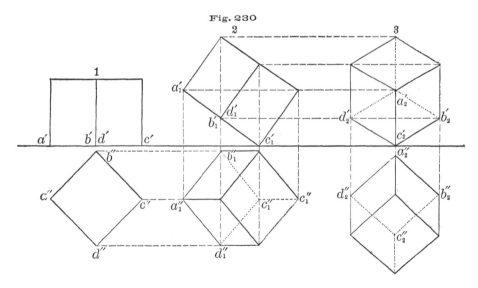

Fig. 230

first. Draw $c_1'a_1'$, making with GL the complement of the given inclination of the axis to H, and complete the vertical projection.

The horizontal projection is determined as in Fig. 229, 2, by the intersections of the ordinates $a_1'a_1''$ and $a''a_1''$, $b_1'b_1''$ and $b''b_1''$, etc., the outermost lines and the upper base being represented in full lines; but $c_1''b_1''$, $c_1''d_1''$ of the lower base and the edge $c_1''c_1''$ are drawn in dotted lines, as they are hidden from view in looking downwards.

(3) Move the cube so placed towards the right and rotate it around a vertical line passing through c_2'' until its axis shall be, for example, in a plane perpendicular to GL (Fig. 230, 3). As the solid by these motions changes its position only to V, each point moving horizontally, in a circle and maintaining its height above H unaltered, therefore the horizontal projection of the third position must be an exact counterpart of that of the second, the sole difference being in the inclination to GL.

The vertical projection will be found, as in Fig. 229, 3, at the intersections of the ordinates $(a_2''a_2',\ a_1'a_2')$, $(b_2''b_2',\ b_1'b_2')$, etc., drawn respectively from the horizontal projection of the third position and the vertical of the second.

227. PROBLEM.—*To determine the projections of an oblique hexagonal pyramid in a doubly oblique position.*

(1) Let Fig. 231, 1 be the surface in its simplest position.

(2) Move the solid to the right and incline the axis until it assumes the angle α to H. During this rotation around $a_1''a_1'$ each point of the surface describes an arc parallel to V, the measure of which is equal to β, the difference between the inclination of the axis to H in its first and in its second positions.

Draw $a_1'd_1'$ (Fig. 231, 2) equal to $a'd'$ and making with GL the angle

Fig. 231

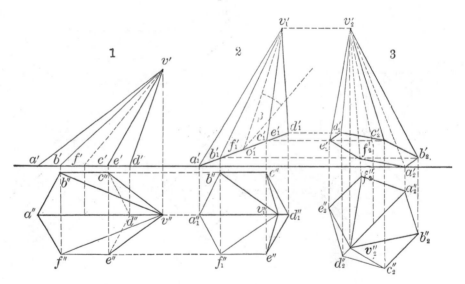

β; set off from a_1' the distances $a_1'b_1'$, $b_1'c_1'$, etc., equal to $a'b'$, $b'c'$, etc., and from o_1' draw the axis $o_1'v_1'$, making the required angle to the base. Complete the new vertical projection by connecting the vertex v_1' with the angular points of the base.

The horizontal projection is found, as in preceding cases, by means of the intersecting ordinates drawn from the second vertical and first horizontal projections. The vertex falling within the base, the entire projection now becomes visible.

(3) Transpose the surface as before, noting that it has not changed its position to *H*, and that, as a consequence, its horizontal projection is precisely the same as that of the second position, the sole difference being in the inclination of the axis to the ground-line.

The vertical projection is determined, as before, by means of the ordinates drawn from the horizontal projection of the third and vertical projection of the second position. The vertex being turned outwards, the entire base becomes visible, while the edges $v_2'a_2'$ and $v_2'f_2'$ are concealed and marked in dotted lines. A test of the accuracy of the construction will be found in the parallelism of those lines which are parallel in space.

228. The projection of an object in a doubly oblique position may be determined without necessarily resorting to the first two auxiliary positions, by the employment of the methods of *rabattement* and supplementary planes. The constructions may likewise be modified by the consideration of the fact that the real lengths of lines in space stand in a fixed proportion to their projections. A scale of reduction may thus be formed and applied to the solution of the third position. In this manner are formed those axonometrical constructions—of which Isometrical Projection is an illustration—the development of which is a special application of Descriptive Geometry.

229. PROBLEM.—*To determine the projections of a right cylinder in a doubly oblique position.*

Fig. 232

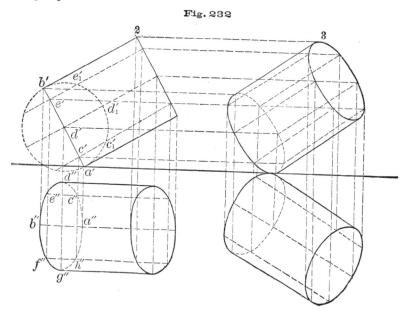

By a *rabattement* of the lower base the first position may be omitted (Fig. 232, 2).

The axis being parallel to V, the vertical projection of the surface is a rectangle the base of which is equal to the diameter of the circle, and the altitude that of the given cylinder. By *rabattement* around a vertical of the plane of the base—preferably ($a'b'$, $a''b''$), which passes through the centre—the circle, projected in its true shape, may be divided into any number of parts, and the intermediate points, c_1', d_1', etc., together with their distances from the axis, readily determine. By counter-rotation the horizontal projections c'', d'', e'', etc., are found by measuring on either side of $a''b''$ distances equal to $c'c_1'$, $d'd_1'$, and $e'e_1'$.

The constructions for the third position do not differ from those already indicated in the preceding cases.

230. PROBLEM.—*To determine the projections of a right cone when the inclination of its axis to the coördinate planes is given.*

The projections may be found without recourse to either of the first two auxiliary projections (Fig. 233).

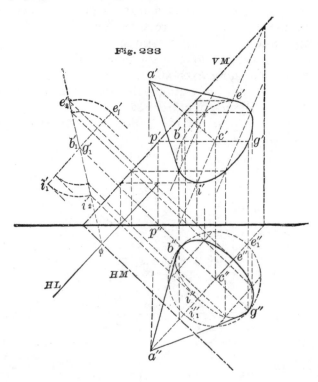

Fig. 233

Let the projections of the axis ($a'c'$, $a''c''$) be determined as in Art. 218. As the base of the cone is at right angles to the axis, the plane passing through the lower extremity of the axis and perpendicular to it coincides with the base. Hence the problem is resolved into the case of a surface which stands upon a given plane, Art. 197.

Pass through the extremity (c', c'') a plane M perpendicular to the line $(a'c', a''c'')$; the horizontal $(c''p'', c'p')$ giving the piercing-point (p', p'') will be sufficient to determine the traces (VM, HM), which are respectively perpendicular to the projections of the axis. By *rabattement* of the plane M around the horizontal the circle of the base is projected horizontally in its true size $e_1''g_1''i_1''b''$.

Assume a new vertical plane L perpendicular to the horizontal $(b'c', b''c'')$; its horizontal trace HL is perpendicular to $b''c''$. Find the new vertical projection $(i_1'e_1')$ of the circle, and the line of greatest declivity $\phi b_1'$, or the intersection between the two planes L and M, and counter-rotate the plane M. During this rotation the plane of the circle remains perpendicular to the new vertical plane, and the projection $(i_1'e_1')$ assumes the position $i_2'e_2'$, each point having described an arc parallel to L and measured by the angle $i_1'b_1'i_2'$. Letting fall ordinates from the different points in this position, the horizontal projections i'', b'', e'', g'', etc., are found in the lines passing through the points i_1'', b'', e_1'', g'', etc., and perpendicular to the horizontal $(b''c'', b'c')$.

The vertical projections i', b', e', g', etc., may be found either by means of the horizontals led through the individual points determined in the horizontal projection or by means of the lines of greatest declivity $e'c'$, $e''c''$, etc.

CHAPTER XI.

SECTIONS.

I. ELEMENTARY SECTIONS.

231. A plane may assume two general positions to a surface: it may either touch or cut it, or, in other words, it may be either a *tangent* or a *secant*. In the first case, as a general rule, the surface lies wholly on one side of the plane, and in the second case it is divided by it.

232. The line in which the cutting plane intersects the surface is termed the *line of intersection*, and is common both to the surface cut and to the cutting plane; hence, whatever the nature of the surface, the line of intersection must be a *plane line*.

233. The character of this line is affected by two considerations: (1) by the nature of the surface itself; (2) by the position of the cutting plane. It is evident that a section of the sphere must be wholly curvilinear and diminish in size as the plane recedes from the centre, while the section of a pyramid must be wholly rectilinear and diminish in size as the plane approaches the vertex.

234. As influenced by the nature of the surface, the section may, therefore, be curved, rectilinear, or a combination of these two; as influenced by the position of the secant plane, the section may vary as the plane passes through a centre, an element, an axis, a diameter, etc.

235. *Longitudinal Sections.*—In every prismatic or cylindric surface, if the secant plane passes through an edge, rectilinear element or axis, it will cut the *longitudinal* section composed of parallel rectilinear elements; and if the surfaces be limited by bases, as in the geometrical solids, the section will be a parallelogram or rectangle.

236. In every pyramidal or conic surface, if the secant plane passes through the vertex, it will cut the longitudinal sections composed of rectilinear elements intersecting in the vertex; and if the surfaces be limited by bases, the section will be a triangle, isosceles,

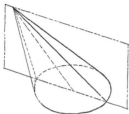

right or scalene as the surface is right or oblique. In the oblique cylinder and cone any longitudinal plane passed through the surface, at right angles to the plane which contains the axis and the line measuring the altitude, will cut the rectangle and isosceles triangle respectively.

237. *Meridian Sections.*—In surfaces of rotation the longitudinal section becomes the *meridian* section, and the secant plane the *meridian plane*. They are determined by passing through the axis a secant plane, which thus coincides with and cuts a generatrix of the surface. Hence the meridian sections of surfaces of rotation *are equal*.

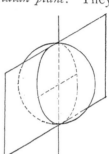

238. It has already been shown that, in generating any surface of rotation, each point of the moving line describes a circle whose plane is perpendicular to the axis and whose centre is in it; hence any section perpendicular to the axis of such a surface must always coincide with one of these circles, termed collectively the *parallels*. Except in the case of cylindric surfaces, these parallels are unequal, the largest being known as the circle of the *equator*, the smallest the circle of the *gorge*.

As every meridian section is cut by a plane which passes through the axis and, hence, through the centres of the parallels, it necessarily divides both parallels and surface into two equal parts.

239. *Perimetrical Sections.*—When the secant plane intersects the consecutive generatrices of a surface, the section thus obtained is termed a *perimetrical* section.

In the cylindric and prismatic surfaces such a section is *closed;* in the pyramidal and conic it may be *open*. Taking an oblique circular cylinder by way of illustration, it will be found that—

(1) The *parallels* are equal to the base.

(2) All sections cut by planes which intersect the surface at an angle α, equal to the inclination of the axis to the base but in the opposite direction, are equal circles, the *sub-contrary sections*.

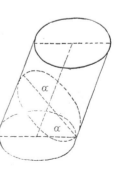

(3) Every section between a parallel and its sub-contrary will be an ellipse with its *transverse* axis equal to the diameter of the base.

(4) Every section beyond the parallel and its sub-contrary will be an ellipse with its *conjugate* axis equal to the diameter of the base.

(5) The smallest section will be cut perpendicular to the axis.

240. In the pyramidal and conic surfaces the section may be open or closed, but all sections cut by parallel planes are similar.

Taking an oblique circular cone by way of illustration, it will be found that the secant plane may intersect:

(1) *All* the rectilinear elements of one nappe only, giving a *closed* section.

(2) All the rectilinear elements of one nappe except that element to which the plane may be assumed parallel, giving an *open* curve.

(3) All the rectilinear elements of both nappes except the two to which the plane may be assumed parallel, giving an *open curve of two branches.*

(4) At an angle equal to the inclination of the axis to the base, but in an opposite direction, giving the *circular sub-contrary sections.*

In the first case the section will always be elliptical, in the second parabolic, in the third hyperbolic, whilst all those sections cut between the second and third will also be hyperbolic.

241. In the greater number of surfaces the sections cut by the perimetrical and longitudinal planes coincide with the generatrices and directrices, and hence are termed the *elementary* sections. In practice they serve as most useful auxiliaries in the determination of lines of intersection.

In prismatic and pyramidal surfaces the edges which mark by their piercing-points on the secant plane the *breaking-points* of the line of intersection are, as a rule, sufficient to determine that line.

In cylindric and conic surfaces the piercing-points of rectilinear elements upon the secant plane indicate the required points in the line of intersection. The parallels, where their employment is readily available, may also be employed to the same end.

In surfaces of rotation the parallels and meridians, in warped surfaces the rectilinear elements, and in other surfaces, as the polyhedrons, the edges and faces, are the auxiliaries employed in the work.

242. As the section is common to the secant plane and to the surface cut, any line of the section must be likewise a line of the plane; hence any point of the line of intersection may be regarded as a point of intersection between a line of the plane and another of the surface. If, then, the surface should be cut, within the possible limits of its intersection, by auxiliary planes, the secant plane will also be cut, and the two lines thus determined lying in the same plane will intersect in one or two points of the line of intersection.

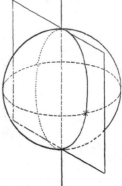

243. An important consideration, therefore, in order to facilitate the graphical constructions is the selection of the position which the auxiliary cutting planes should be made to assume so as to afford the simplest

sections of the surface and the most available arrangement for projection.

On the surface itself the secant planes should be:

(1) For prismatic and cylindric surfaces parallel to the axes.

(2) For pyramidal and conic surfaces passed through the vertices.

(3) For surfaces of rotation made to cut the parallels.

244. PROBLEM.—*Through a point on the surface of a cone to cut a longitudinal and a parallel section.*

The longitudinal plane passes through the vertex (v', v'') and, hence,
through the line joining the given point (a', a'') with the vertex (Fig.
234).

From these data an indefinite number of sections may be cut, all of

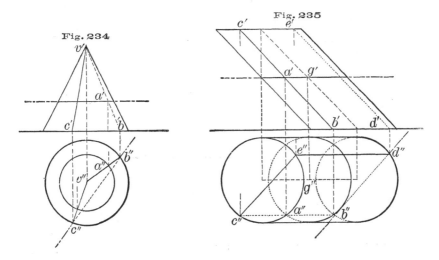

Fig. 234

Fig. 235

which are determined by planes which pass through the foot (b', b'')
of the line, intersect the base in any line ($b''c''$, $b'c'$), and cut ($c'v'$, $c''v''$),
thus completing the triangle. Only one parallel can be cut, which,
being parallel to H, is projected thereon in a circle which contains
the given point.

The point being common to both cutting planes lies at the intersection (a'', a') of their respective sections.

245. PROBLEM.—*Through a point on the surface of an oblique cylinder to
cut a longitudinal and a parallel section.*

Through the given point, of which the vertical projection a' (Fig.
235) only is given, lead an element ($b'c'$, $b''c''$) of the surface and determine the horizontal projection a''. As an indefinite number of secant
planes may be passed through this element, select at will any second
element ($e'd'$, $e''d''$), and complete the section by connecting the extremities of the elements.

But one parallel can be cut, whose centre ($g'g''$) lies in the axis and
whose projections contain those of the given point.

246. PROBLEM.—*To divide a horizontal prism into two equal parts by a longitudinal plane which makes a given angle with the horizontal plane; and to remove the front section of the surface.*

Let the prism be octagonal (Fig. 236).

Assume a new ground-line G_1L_1 perpendicular to the primitive ground-line, and determine the new vertical projection of the surface. As the secant plane is perpendicular to the new vertical plane, the projection of the section falls in any line $m_1'n_1'$. From the new vertical projections the primitive projections of both surface and section are readily found according to Art. 154.

The broken lines represent the portion of the surface removed. When the section is exposed to view by the removal of any part of the surface, the fact is indicated by *shading the section.*

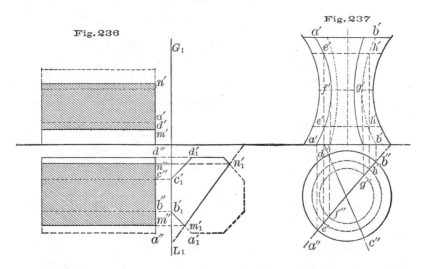

Fig. 236   Fig. 237

247. PROBLEM.—*To cut the meridians on a surface of rotation.*

As the surface may always be brought by methods already demonstrated into a position in which the axis shall be perpendicular to H, assume that position as in Fig. 237.

The secant planes are then perpendicular to H, and the sections projected on it in right lines, as $a''b''$, $c''d''$, etc. The auxiliary planes which cut the parallels on the surface (Art. 242) and the lines on the secant planes are thus horizontal, the points of intersection a'', e'', f'', etc., between these two sets of lines marking the required points in the line of intersection.

The vertical projections are obtained by drawing ordinates from these points to the auxiliary circles in which they are contained. The meridians pass through them.

II. OBLIQUE SECTIONS.

248. Where a surface is cut by a plane which does not coincide with the primitive elements, the section is oblique. Analytical Geometry determines the character of such sections, and the results of mathematical investigations have established the following facts with regard to them:

(1) The cylinder gives *ellipses.*

(2) The cone, *ellipses, parabolas, hyperbolas.*

(3) In surfaces of rotation—

    (*a*) The ellipsoid gives *ellipses.*

    (*b*) The paraboloid, *ellipses* and *parabolas.*

    (*c*) The hyperboloid, *ellipses, parabolas* and *hyperbolas.*

(4) In surfaces of the second order any plane section is limited by a line of the second order.

(5) In surfaces of the second order parallel sections are similar.

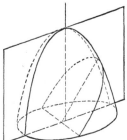

249. The secant plane always cuts a surface of rotation in a symmetrical figure, the transverse axis of which is the line of intersection between the secant plane and the meridian plane at right angles to it. The point or points in which this axis pierces the surface will be the vertex or vertices of the section cut.

250. In determining the perimeter of the oblique section · the elements, edges or other auxiliary lines of the surface may be employed; the practical requirements of the case demand, however, that the choice shall lie with those lines which render the constructions most simple in character.

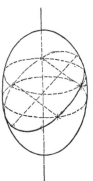

251. A system of auxiliary planes which cut a series of auxiliary lines on the surface will also intersect the plane of the oblique section in right lines, which are secants to that section. These secants pierce the surface in the points in which the oblique plane crosses the auxiliary lines traced upon the surface.

252. PROBLEM.—*To intersect a given prism by a plane perpendicular to the vertical plane.*

Let the prism be given as in Fig. 238.

By the conditions of the problem the vertical trace of the secant plane may be any line *VM*, while its horizontal trace *HM* must be perpendicular to *GL*. The section lying in a plane perpendicular to *V* is projected on that plane in a line *a'c'* in the vertical trace; and as the entire prism is projected horizontally in the perimeter of the base, it is evident that any part thereof, as the section, must likewise be projected therein.

The true size of the section may be determined by rotating the plane *M* around either trace or around one of its verticals, preferably that which intersects the axis. The latter construction is usefully employed where the space is restricted.

Fig. 238

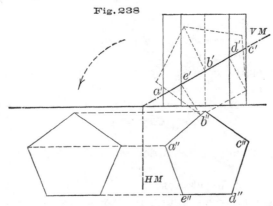

253. PROBLEM.—*To intersect an oblique pyramid by a plane perpendicular to the horizontal plane.*

Fig. 239

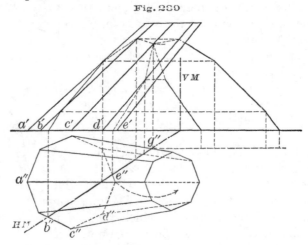

Let the pyramid be given as in Fig. 239.

The secant plane (*VM*, *HM*) being perpendicular to *H*, the section it cuts is projected in the horizontal trace, *HM*, in the line $b''g''$; the points in which the trace cuts the horizontal projections of the edges are projections of the breaking points of the line of intersection. Hence the vertical projections are found by drawing the ordinates from these points to the vertical projections of the edges in which they lie.

The true size of the section may be determined by *rabattement*, as in the preceding case, or around a vertical of the plane passing through the point (g', g'').

254. PROBLEM.—*To intersect a right cylinder by a given plane.*

Let the cylinder be given as in Fig. 240, and let *VM*, *HM* be the traces of the given plane.

Cut the plane and the surface by a series of auxiliary planes (Art. 251), the surface in rectilinear elements and the plane in verticals. In accordance with these conditions the auxiliary planes must be parallel to *V*, giving the traces $c''k''$, $b''l''$, $a''h''$, etc., and cutting the cylinder in elements the horizontal projections of which fall in c'', a'', b'', etc. Determine the vertical projections of these elements $e'e'$, $d'd'$, etc., and of the verticals $c'k'$, $b'l'$, $a'h'$, etc., and their points of intersection mark points in the line of intersection sought. Thus, $c'c'$ the element intersects $k'c'$ the vertical at the point c', $b'b'$ intersects $b'l'$ at b', and similarly for intermediate points of the section $e'f'a'b'c'd'$.

The plane of the upper base being horizontal cuts the secant plane *M* in a horizontal ($e''f''$, $e'f'$), which completes the limits of the section.

To determine the true size of the section, bring it by *rabattement* around one of the horizontals, as $e''f''$ of the secant plane.

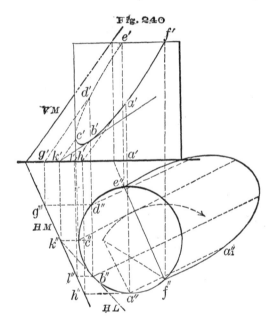

Fig. 240

255. PROBLEM.—*To intersect a right cone by a given plane.*

Let the cone be given as in Fig. 241, and let the plane *M* be perpendicular to *V*.

Pass through the surfaces a series of longitudinal cutting planes at right angles to *V*. Each plane cuts on the cone two rectilinear elements, as ($b''v''$, $b'v'$) and ($h''v''$, $h'v'$), which coincide in the vertical projections, and a line ($h''b''$, $h'b'$) on the secant plane which intersects the elements in the points b and h respectively. A succession of points may thus be obtained through which the required section *abcdefgh* passes. For

the point (c', c''), however, as the auxiliary sections coincide in projection and, hence determine no point of intersection, a parallel instead of

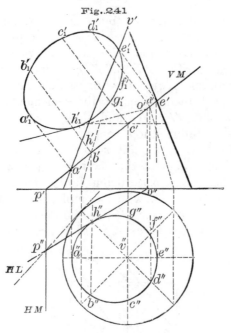

a longitudinal section has been employed.

To determine the section in its true size, revolve the secant plane
around its vertical trace *VM*.

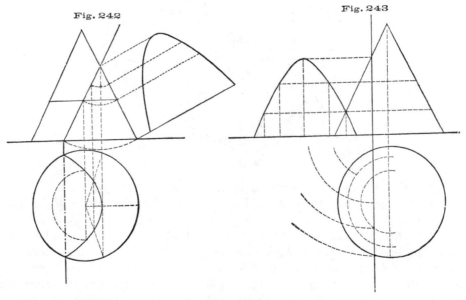

Fig. 242 exhibits the case in which the secant plane has been assumed
to be parallel to an element of the cone. The section which is the

parabola may be found by constructions similar to those of the preceding example.

Should the secant plane be taken parallel to the axis (Fig. 243), or in any position intermediate between this and that immediately preceding, the *hyperbola* will be the section cut, since both nappes of the surfaces will be intersected. In these cases the parallels will prove the most available auxiliaries.

256. PROBLEM.—*To intersect a surface of revolution by a given plane.*

Let the surface be given as in Fig. 244, and let (*VM, HM*) be the given plane.

By Art. 249 the axis of the section will be determined by finding the intersection ($p'o'$, $p''o''$) between a meridian plane L and the given plane, the two being at right angles to each other. The points in which this line pierces the surface are the vertices of the curve of intersection as well as the *highest* and *lowest* points thereof.

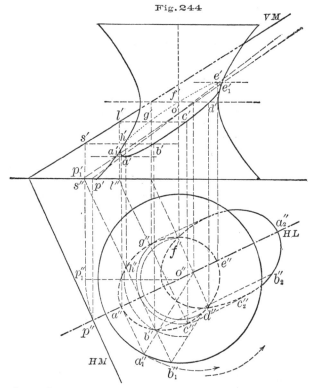

Fig. 244

To determine these points revolve the meridian plane around the axis of the surface until it becomes parallel to V; the line ($p'o'$, $p''o''$) then assumes the position ($p_1''o''$, $p_1'o'$), in which a_1' and e_1' are the vertices revolved. By counter-rotation (a', a'') and (e', e'') indicate the vertices in the required position.

Now, by a series of horizontal auxiliary planes between these two limiting points, cut parallels on the surface and horizontals on the secant plane *M*; the intersections of these two fix a succession of points, *a, b, c, . . .* through which the required section passes. Thus the horizontal $(s''b'', s'b')$ cuts the circle $(b''h'', b'h')$ in the points (b'', b') (h'', h'), the horizontal $(l''g'', l'g')$ cuts the circle $(c''g'', c'g')$ in the points (c'', c') (g'', g'), etc. To determine the section in its true size, bring it by *rabattement* around a horizontal, as $(f'd', f''d'')$.

257. PROBLEM.—*To intersect a pyramid by a given plane.*

Let the pyramid be given as in Fig. 245, and let (VM, HM) be the given plane.

Pass through the edges a series of auxiliary cutting planes perpendicular to *V*, intersecting the surface in the edges and the secant plane, *M*, in a series of lines; the points in which these lines intersect the edges are the piercing-points of the latter on the secant plane or the breaking-points of the line of intersection sought.

As the auxiliary planes all pass through the axis, their intersections

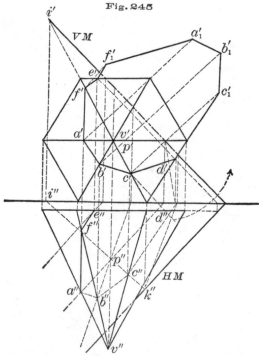

Fig. 245

with the secant plane *M* must pass through the piercing-point (p', p'') of the axis on that plane. Thus the auxiliary plane passing through the edges $(v'f', v''f'')$ and $(v'c', v''c'')$ cuts the secant plane in a line whose horizontal projection $i''k''$ intersects the edges in the points (f', f'') and (c', c''). Similar constructions determine the points (a', a'') and (b', b''), whilst the points (d', d'') and (e', e'') may be found by passing auxiliary planes through the sides of the base in which they lie.

The true size of the section may be determined by *rabattement* around the line $(d'e', d''e'')$.

258. PROBLEM.—*To intersect an oblique cylinder by a plane at right angles to the axis.*

Let the cylinder be given as in Fig. 246, the surface extending indefinitely upwards. By the conditions of the case the traces (*HM, VM*) of the plane must be perpendicular to the projections of the axis of the same name. The section thus cut is termed a *right section* of the surface.

Pass a series of *vertical* longitudinal planes cutting elements upon the surface and parallel right lines upon the secant plane *M*. Thus, the plane passing through the elements whose feet are a'' and e'' cuts the plane *M* in a line whose horizontal projection coincides with that of the elements. One point of its vertical projection may be immediately found at o', and the second, n', may be determined by means of a hori-

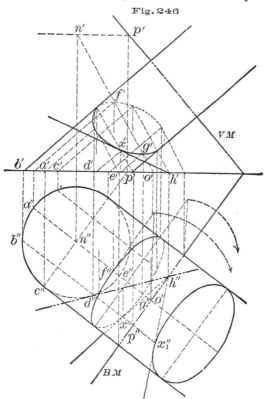

Fig. 246

zontal led through any point p' of the vertical trace of the plane. Hence the line to which the vertical projections of all lines cut on the plane *M* by the auxiliary planes must be parallel passes through o' and n'. The intersections of ($o'n'$, $o''n''$) with the elements passing through (a', a'') and (e', e'') determine the points (f', f'') and (g', g'') of the right section sought. By a similar procedure a series of rectilinear elements and of lines parallel to ($o'n'$, $o''n''$) may. be obtained, whose intersections mark points in the vertical projection of the curve of the section, from which the horizontal projections may readily be determined by means of ordinates.

To determine the section in its true size, bring the plane *M* by *rabattement* around its horizontal trace *HM*.

259. PROBLEM.—*To pass a plane through a given line, and to cut a great circle on a sphere.*

Let the sphere be given as in Fig. 247, and let *L* be the given line.

As every plane cutting a great circle must pass through the middle of the sphere, from any point (o', o'') of the given line lead a line through the middle (c', c'') and determine the traces (*VM*, *HM*) of the plane passing through these two lines (Art. 85). The meridian plane (*VN*, *HN*) drawn perpendicular to the plane thus found cuts the transverse axis of the section whose extremities mark the highest and lowest points thereof (Art. 249).

Fig. 247

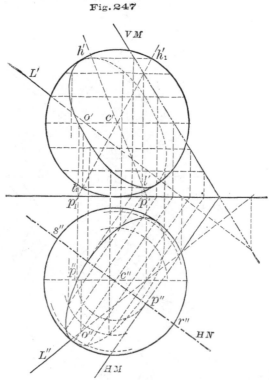

Thus the meridian plane *N* cuts the line ($p''c''$, $p'c'$) on the plane *M*, and on the sphere a great circle whose horizontal projection is $r''s''$. Rotate the plane *N* around the vertical axis of the sphere until it is brought parallel to *V*, when the line ($p''c''$, $p'c'$) assumes the position ($p_1''c''$, $p_1'c'$) and determines the points h_1' and l_1', the resolved positions of the vertical projections of the highest and lowest points respectively of the section sought.

By counter-rotation h_1' falls at h', and l_1' at l'. Now by a series of auxiliary horizontal planes between these two points cut parallels on the sphere and horizontals on the secant plane *M*; their intersections mark points of the great circle required.

CHAPTER XII.

INTERSECTIONS.

I. ELEMENTARY INTERSECTIONS.

260. When the distance between two surfaces is reduced to a minimum they merely touch or are *tangent* to each other, but if this distance be measured by a negative quantity they *intersect*. In such a position they have *a line in common*, termed the *line of intersection*.

261. This line of intersection being common to the two surfaces, it follows that each of its points is likewise common; hence, in determining points which are common to two surfaces the line of intersection will be determined.

262. The character of the line of intersection is affected by three considerations: the nature of the surfaces themselves, their relative positions and their extension beyond the initial line of intersection. Thus it is evident that the rectilinear surfaces can only intersect in right lines, double-curved surfaces in curved lines, whilst a polyhedron and a curved surface may present a line which is partly right **and partly curved.**

263. It is to be noted that, other things being equal, a change in position materially modifies the character of the line of intersection. Thus a pyramid and a cone whose vertices coincide intersect in a triangle, but for all other positions the plane faces of the pyramid cut, as a rule, perimetrical sections on the surface of the cone.

264. The variations in the line of intersection which arise from the prolongation of the surfaces beyond their initial intersection may be illustrated by means of the cylinders of rotation.

Assuming the axes of two equal cylinders to intersect, it is found that:

(1) Should the surfaces stop at their first contact, the line of intersection will be an ellipse or oblique section of either surface, and that the smaller the angle a between the axes the greater the transverse axis of the ellipse. In such a case the surfaces are said to intersect by *meeting*.

(2) Should *either* surface be extended beyond the first line of contact, the intersection assumes the form of a wedge, whose two faces are semi-ellipses which respectively increase

and diminish with the angle between the axes. In such a case the surfaces are said to intersect by *penetrating.*

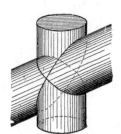

(3) Should both surfaces be extended beyond the first line of contact, the line of intersection is composed of two ellipses. In such a case the surfaces are said to intersect by *crossing.*

265. *Requirements for Elementary Intersections.*—Two prisms or two cylinders, or a prism and a cylinder, will intersect in rectilinear elements, provided their axes or edges are parallel.

Two cones or pyramids, or a pyramid and a cone, will intersect in rectilinear elements, provided their vertices coincide.

Two pyramids, or a pyramid and a prism, whose parallel sections are similar and whose homologous sides are parallel will, should the axes coincide, intersect in one of the parallels.

Surfaces of revolution whose axes coincide intersect in parallels or circles.

266. PROBLEM.—*To dispose a right prism and right cylinder so as to intersect in rectilinear elements.*

Let the axes be made parallel (Fig. 248); then the faces of the prism, being parallel to the axis of the cylinder, cut that surface in the elements $(a'', a'a')$ and $(b'', b'b')$.

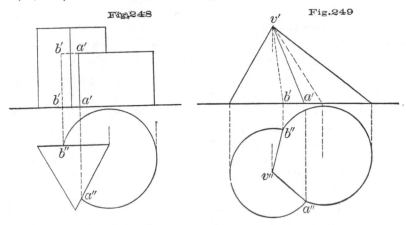

267. PROBLEM.—*To dispose two cones so as to intersect in rectilinear elements.*

Let the vertices coincide (Fig. 249); then the elements $(v'a', v''a'')$ and $(v'b', v''b'')$, passing through the vertices and the points of intersection of the bases, are common to the two surfaces, and hence are their line of intersection.

268. PROBLEM.—*To dispose a right pyramid and right prism whose respective parallels are similar, so as to intersect in a parallel.*

Let the axes coincide (Fig. 250) and the sides of the bases be parallel; then each face of the prism cuts a face of the pyramid in a line which is parallel to a side of its base, while the edges, lying four and four in the same vertical planes, intersect in the breaking-points of the parallel of intersection.

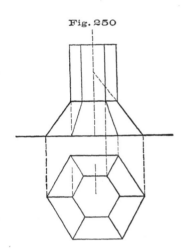

Fig. 250

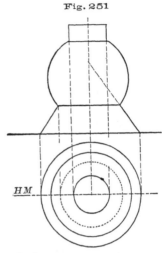

Fig. 251

269. PROBLEM.—*To dispose surfaces of revolution so as to intersect in parallels.*

Let the axes coincide (Fig. 251); then any meridian plane *M*, as for example that parallel to *V*, determines points on the apparent contour through which the required parallels pass.

270. PROBLEM.—*To determine the circle of intersection between two spheres.*

Let the spheres be given as in Fig. 252.

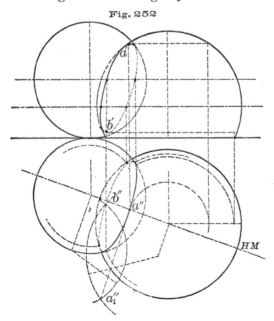

Fig. 252

Pass a vertical meridian plane, *M*, through the centres of the two surfaces, and determine, by the *rabattement* of the plane around its horizontal trace *HM*, the highest and lowest points, (*a''*, *a'*) and (*b''*, *b'*) respectively, of the line of intersection. Cut any number of auxiliary parallels between these points by means of horizontal planes; they intersect, two and two, in points of the line of intersection sought.

II. PLANE LINES OF INTERSECTION.

271. If the generatrices of two surfaces be made to follow a common directrix, the surfaces thus generated will intersect each other in this line.

272. The character of the plane line of intersection will be affected by the following general limitations: first, that the surfaces admit of *identical* sections; second, that they are *collinearly* disposed—that is, have their homologous elements in the same relative position to a median plane; third, that the surfaces simply *meet*.

273. *Requirements for Plane Intersections.*—Two *equal* cylinders of revolution—that is, cylinders whose parallels are *identical*—are collinearly disposed when their axes are either parallel or intersect, thus affording the longitudinal or perimetrical intersections respectively.

Two equal prisms are similarly disposed when, their axes being parallel or made to intersect, a longitudinal plane passing through these axes divides the surfaces identically. The intersection is some polygon which can be cut on both surfaces.

Two equal cones—that is, cones whose parallels at the *same distances* from the vertices are identical—are collinearly disposed when, their axes being parallel, a plane cutting equal parallels is at the same distance from each vertex; or when the axes are made to intersect in a point whose distance from the vertices is the same. The intersection under these conditions is either the parabola, hyperbola, or ellipse.

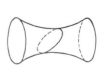

Two equal pyramids will be collinearly disposed when, in addition to the conditions imposed upon the cones, the bases have identical positions with reference to a longitudinal plane which passes through the axes.

Any two equal surfaces whose homologous edges or elements intersect afford plane lines of intersection.

Any two surfaces upon which identical sections can be cut may be so disposed as to intersect in that section.

274. Where surfaces, such as some of the regular geometrical and surfaces of revolution, can be cut in bi-symmetrical sections, an intersection may be effected by reversing the section on either side of the cutting plane.

275. By Analytical Geometry it has been demonstrated that when

two cylinders, or two cones, or a cone and a cylinder, are made to circumscribe a sphere in common, they intersect in a plane line of intersection.

276. PROBLEM.—*To find the line of intersection between two equal cylinders
whose axes intersect.*

Let the cylinders be given as in Fig. 253, in which the rectilinear
elements, having the same position to the coördinate planes, give projections which make equal angles with the ground-line.

Pass through the two surfaces a series of auxiliary planes cutting
rectilinear elements upon each; these planes are parallel to the longitudinal plane containing the axes, and give horizontal traces parallel to
$a''i''$. Find the points in which each pair of elements so cut intersect
each other. Thus the plane whose horizontal trace is $d''g''$ cuts on each
surface an element whose feet are in the points d'' and g'' respectively.
The intersection c' of their vertical projections immediately determines a
point in the line of intersection sought, from which the horizontal projection c'' may be found by means of an ordinate.

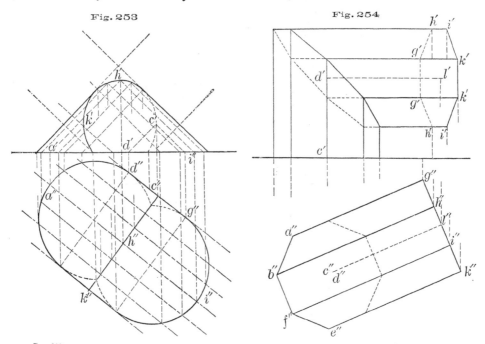

Fig. 253   Fig. 254

In like manner each auxiliary plane determines a point which, together
with the highest point (h', h'') found by the plane passing through the
axes and the lowest points found at the intersection of the bases, fixes
the position of the line of intersection sought.

The plane of the curve being perpendicular to *H*, $c''k''$ is its horizontal projection.

277. PROBLEM.—*To find the line of intersection between two equal prisms
whose axes intersect at right angles.*

Let the axes *cd* and *dl* be given as in Fig. 254, and let $a''b''f'' \ldots$
be the horizontal projection of the base of the vertical prism.

Since the base of the second prism is perpendicular to *H*, assume
any line, $g''k''$, perpendicular to $d''l''$ as its horizontal projection. By
rabattement and counter-rotation around the horizontal passing through
(l', l'') determine the vertical projection $g'h'i'k'g'$, when the homologous
edges of the surfaces immediately intersect in points of the line of intersection whose horizontal projection falls in the perimeter of the base
of the vertical prism.

278. PROBLEM.—*To find the line of intersection between two equal pyramids whose axes intersect at right angles.*

Let the axes *oc* and *oi* be given as in Fig. 255, and let $a''b''d''e'' \ldots$
be the horizontal projection of the smaller base of the vertical pyramid.

As the point of intersection of the axes (o', o'') must be at the same
distance from the vertices (Art. 273), make $o''i''$ equal to $c'o'$, and $g''k''$
equal to $b'f'$. With (o', o'') as a centre, find the projections of any

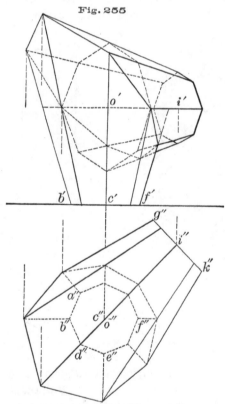

Fig. 255

larger base parallel to *H* of the vertical pyramid, and by rotation of
the same around a horizontal passing through (o', o'') determine its
vertical projection for the second pyramid. The bases of the respective
pyramids, being collinearly disposed with reference to the plane of the
axes, the edges immediately intersect in points of the required line of
intersection.

279. PROBLEM.—*To find the line of intersection between two equal cones.*

Let a vertical cone be given as in Fig. 256, and let the horizontal axis of the second cone intersect that of the first in the point (f', f'').

Since the axes of the surfaces intersect at the same distance from the vertices (Art. 273), make $f''u''$ equal to $f'v'$, and determine u'. The simplest auxiliary sections are those cut by planes passing through the vertices or the line $(v'u', v''u'')$ which joins them. Such a series of planes cut rectilinear elements on the surfaces and give traces which contain the piercing-points (o', o'') and (p', p'') in common.

Fig. 256

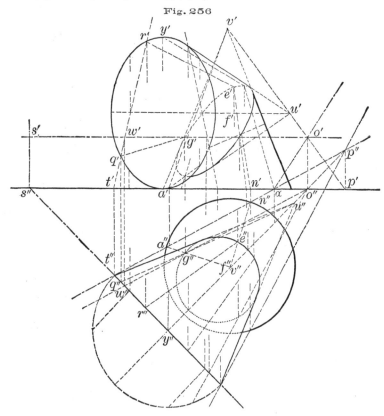

Thus the plane passing through (a', a'') gives the traces $a''\alpha$ and $\alpha o'$, and cuts the vertical cone in the elements $(a''v'', a'v')$ and $(n''v'', n'v')$. It likewise cuts the plane of the base of the horizontal cone in a line whose points of intersection with that base are the feet of the elements cut on the cone. One point of the line lies at the intersection (t', t'') of the horizontal traces, the other point (w'', w') at the intersection of the horizontals $(s''w'', s'w')$ and $(o''w'', o'w')$, found by means of the auxiliary horizontal plane whose vertical trace is $o's'$. Hence, drawing the line $(t'w', t''w'')$, (q', q'') and (r', r'') are the feet of the elements sought, and (g', g'') (e', e'') points in the line of intersection between the two cones. By a repetition of this method a sufficient number of points may be found to determine the entire line.

III. INTERSECTIONS IN WARPED LINES OF INTERSECTION.

280. For all other positions not indicated in the preceding two chapters surfaces will, in general, intersect in lines which cannot be contained in a single plane. As already shown, surfaces which by simply meeting would present plane lines of intersection will when extended give lines whose branches lie in two distinct planes.

Disregarding these latter cases, which may be considered as special illustrations of the second division, the class of intersections now under discussion may be grouped under the heading of *warped* lines of intersection.

281. The warped lines of intersection may be arranged in general classes according to the nature of the surfaces which are made to intersect.

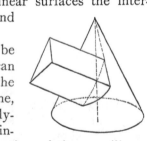

(1) Polyhedral surfaces which meet always intersect in rectilinear intersections, the separate faces determining the various plane branches thereof.

(2) Polyhedral and ruled surfaces which meet always intersect in a broken curved or a mixed line of intersection composed of rectilinear and curvilinear elements.

(3) Double-curved surfaces intersect in double-curved lines of intersection of one or two branches.

282. With a system of polyhedrons three classes of intersections may be noted:

(1) Where the edges of one surface intersect those of the other.

(2) Where the faces of one intersect the edges of the other.

(3) Where the faces of one intersect the edges and faces of the other.

283. With a system of polyhedral and curvilinear surfaces the intersections may be grouped according as the second surface is ruled or double-curved.

(1) If ruled, the line of intersection may be composed of any or all the elements which can be cut on such a surface. Thus on the cone the line may be partly a right line, when one face of the polyhedron coincides with a rectilinear element, and partly any portion of the curvilinear elements.

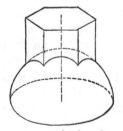

(2) If double-curved, the line of intersection must be wholly curved, each face of the polyhedron cutting a curve whose extremities are the breaking-points of the line of intersection.

284. With a system of two curvilinear surfaces the lines of intersection may be classified as the surfaces are both ruled, or one ruled and one double-curved, or both double-curved.

In all of these cases the line is curvilinear, and presents one or two open or closed branches symmetrically or unsymmetrically disposed, as the relative position of the two surfaces varies.

285. PROBLEM.—*To find the line of intersection between a right pyramid and a right prism.*

Let the surfaces be given as in Fig. 257.

As the breaking-points of the line of intersection are the piercing-points of the edges on either surface, pass through the axes and edges a series of auxiliary cutting planes. The right elements cut on the surfaces, and the edges lying in the same planes, respectively intersect in the breaking-points required.

Thus the auxiliary plane passing through the edge ($f'v'$, $f''v''$) cuts

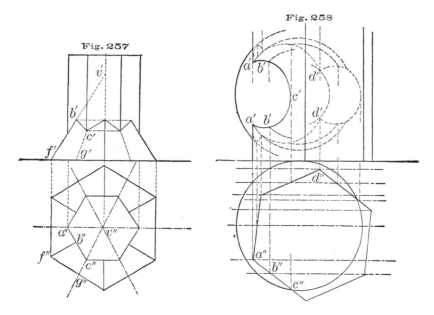

Fig. 257   Fig. 258

on the prism the element whose horizontal projection falls in b'', thus determining (b', b'') of the line of intersection. In like manner the plane passing through the edge of the prism whose horizontal projection is c'' cuts on the pyramid the element ($g'v'$, $g''v''$), thus determining the point (c'', c'); and as the two surfaces are symmetrically disposed to each other, the remaining points may be found by means of horizontal lines drawn through b' and c'.

286. PROBLEM.—*To find the line of intersection between a right prism and a sphere.*

The intersection will be wholly curvilinear, each face of the prism cutting upon the sphere arcs of circles whose extremities are the piercing-points of the edges upon that surface.

Let the surfaces be given as in Fig. 258.

Pass through the edges auxiliary planes parallel to V, cutting on the sphere circles parallel to that plane, and determine the points of intersection a', a'', d', d'', etc., between these circles and the edges. Intermediate points, as (b', b'') and (c', c''), may be found by means of similar cutting planes, in sufficient number to mark the entire line of intersection.

287. PROBLEM.—*To find the line of intersection between two oblique prisms.* Let the surfaces be given as in Fig. 259.

The simplest auxiliary planes are those which, being parallel to the axes of the two prisms, cut rectilinear elements on each.

To determine the position of such a series of planes, from any point (a', a'') of one axis draw a parallel ($a'b'$, $a''b''$) to the other, and through their piercing-points, c'' and b'', draw the horizontal trace of the plane which contains them. The horizontal traces of all the auxiliary cutting planes are parallel to $c''b''$.

Thus the plane passing through the edge whose foot is d'' gives a horizontal trace $d''g''$ and cuts the second prism in the edge whose foot is g''; the point (o', o'') common to the two edges is a point in the line of intersection sought.

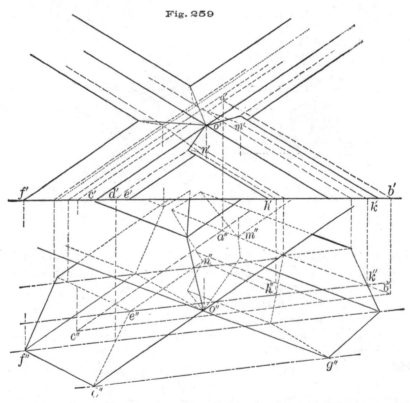

Fig. 259

Again, the plane passing through the edge whose foot is e'' cuts on the second prism the elements whose feet are h'' and k''; the points (m', m'') and (n', n'') in which the edge crosses these elements are likewise points in the line sought. The repetition of this method for the other edges of *both* surfaces determines the breaking-points necessary to describe the entire line of intersection.

288. PROBLEM.—*To find the line of intersection between a cylinder and a cone.*

Let the surfaces be given as in Fig. 260.

Through the two surfaces pass a series of secant planes cutting right

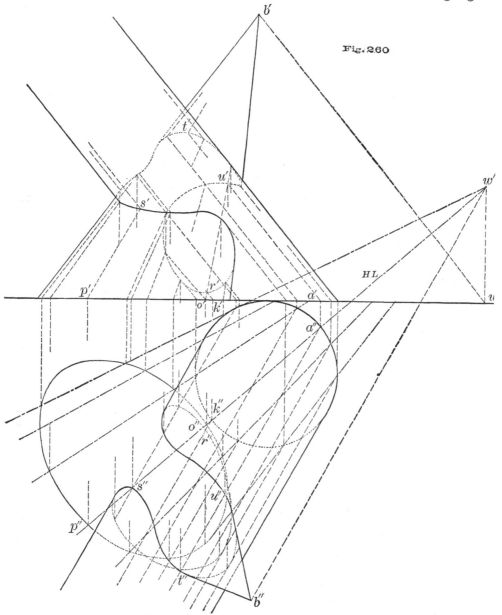

Fig. 260

elements on each. To effect this lead through the vertex (*b'*, *b''*) of the cone a line (*b'w'*, *b''w''*) parallel to the axis of the cylinder, and find its horizontal piercing-point, w''. Any secant plane passing through this line will cut the desired elements.

Assume, then, the horizontal traces of such a series of planes, all of which pass through the piercing-point w'', and determine the elements in which they cut the respective surfaces; their intersections are points in the line of intersection sought. Thus, the plane L cuts the cylinder in two elements whose feet are a'' and k'', and the cone in two elements whose feet are o'' and p''; the points (r', r'') (s', s'') (t', t'') and (u', u'') in which they intersect are points of the line of intersection. In a similar way a sufficient number of points may be found to determine the entire line of intersection.

Note.—In determining the respective portions which are either seen or concealed on the combined projections of the intersecting surfaces it is to be observed that—

(1) The outline of the united projections is always visible.

(2) The line of intersection being common to both surfaces, the vertical projection of that line is visible only so far as it lies on the *front portion* of either surface.

(3) The horizontal projection of that line is visible only so far as it falls on the *upper portion* of either surface.

(4) In passing from the part seen to that which is hidden, the visible limits always terminate at the apparent contour of either surface.

289. PROBLEM.—*To find the intersection between a cylinder and a sphere, the sphere tangent interiorly to the cylinder.*

Let the cylinder be given as in Fig. 261, and let the centre of the sphere lie in the meridian plane passing through the element whose foot is (a'', a').

Determine the centre of the sphere by *rabattement* of the meridian plane around HL, carrying with it the element and the meridian section of the sphere tangent to it. By counter-rotation the *rabattement* of the centre c_1'', the tangent point t_1'' and the highest point h_1'' of the line of intersection fall respectively in c'', t'' and h''. From c'' as a centre describe the horizontal projection of the sphere, and from c'—determined by measuring the distance $c''c_1''$ above H—the vertical projection of the same.

The parallels of the two surfaces being the simplest constructive elements, pass between the highest and lowest points, h' and the bases, respectively, a series of horizontal cutting planes; the points in which the sections intersect are points in the line of intersection sought.

Thus the plane M cuts on the sphere the circle whose radius is $(i'b', i''b'')$, and on the cylinder the circle whose centre is (e', e''), which intersect each other in the points (d', d'') and (g', g''). Similar constructions for the other secant planes mark a sufficient number of points to determine the entire curve of intersection. Should the sphere be entire, the curve thus determined is composed of two closed branches which cross at the point of tangency.

290. PROBLEM.—*To find the intersection between two surfaces of revolution whose axes lie in a common meridian plane.*

Let the two surfaces be given as in Fig. 262, the axes intersecting in the point (p', p'').

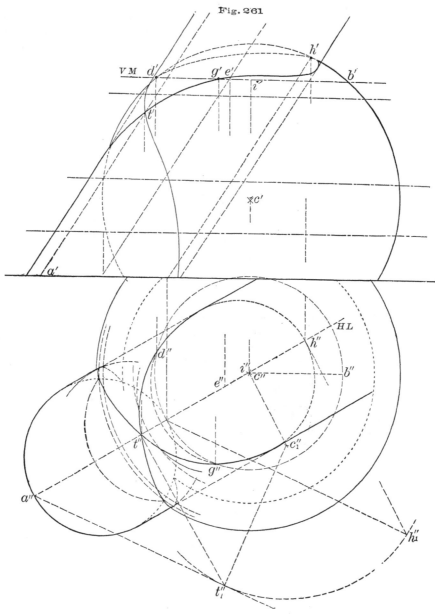

Fig. 261

In such a case plane auxiliary sections do not present simple constructive elements; but if the point (p', p'') be assumed as the centre of a series of auxiliary spheres, then each surface will be intersected in a circle the planes of which are perpendicular to the respective axes. As the circles so cut are common to the spheres, they intersect in points of the line of intersection sought.

Thus the first auxiliary sphere intersects the ellipsoid in the circle whose vertical projection is $a'b'$, and the paraboloid in the circle whose vertical projection is $c'd'$; the two points in which these circles intersect coincide in e' and are points in the line of intersection. The points (f', f'') (g', g'') found by means of additional spheres, together with the highest point (h', h'') and the lowest point (l', l''), are sufficient to determine the entire line.

Fig. 262

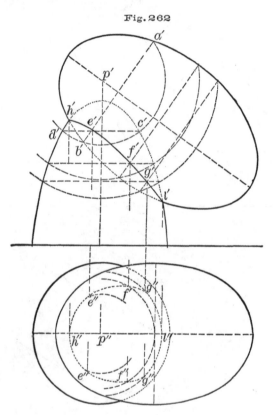

291. PROBLEM.—*To find the intersection between two surfaces of revolution whose axes do not lie in the same plane.*

Let the surfaces be given as in Fig. 263, the axes of both surfaces being parallel to *V*.

In this position it is possible neither to employ secant planes to cut the surfaces in parallels, nor auxiliary spheres to cut them in circles. The simplest constructive method which remains is the use of horizontal planes giving on the first surface parallels, and on the second sections whose horizontal projections must be determined by points.

Thus, let *VL* be the vertical trace of such a secant plane, cutting the first surface in a circle ($a'b'$, a'' . . . b''), and the second in a section whose vertical projection is the line $c'd'$ and whose horizontal projection may be found by the following construction:

Pass through the surface a series of planes perpendicular to the axis $(e'f', e''f'')$. Let $x'y'$ be the vertical projection of a circle cut by such a plane; then the point z' in which it intersects $c'd'$ is a point of that auxiliary section, and by the *rabattement* of the circle $x'y'$ around

Fig. 263

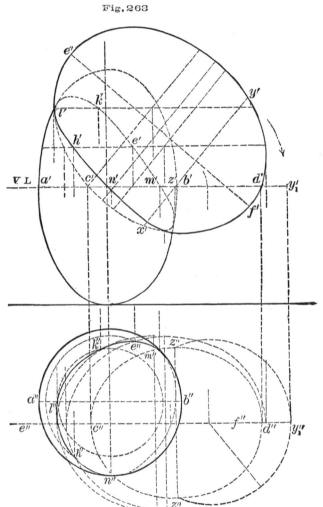

the horizontal projected vertically in z' the horizontal projection z'' may be determined. In a similar way any number of points of the auxiliary section $(c'z'd', c''z''d'')$ may be obtained. The curve thus found, and the circle whose vertical projection is $a'b'$, lying in the same plane, intersect in two points (m', m'') and (n', n'') of the line of intersection sought.

By repetition of this method a series of points may be found through which the entire line of intersection *hiklmn* passes.

CHAPTER XIII.

TANGENTS AND NORMALS.

I. GENERAL CONSIDERATIONS.

292. Any right line lying in the plane of a curve and intersecting it in two points *a* and *b*, which are not consecutive (Fig. 264) or are separated by a finite distance, is a *secant* to that curve. Should such a line be turned around *a* as a pivot, the point *b* gradually shortens its distance from *a* until it reaches a position in which it becomes consec-

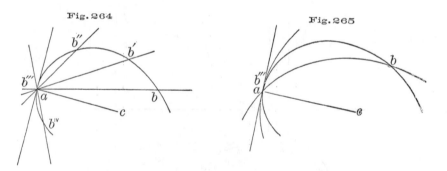

Fig. 264   Fig. 265

utive with it. In this position *ab'''* no longer intersects the curve, and is said to be *tangent* to it. Should the motion be continued, the points again begin to separate and the line becomes a secant.

Hence the tangent may be defined to be the limit of a secant drawn from one point of a curve to another point of that curve at an infinitely small distance apart.

The point *a* is termed the *point of contact*, and the line *ac*, drawn at right angles to the tangent *ab'''*, the *normal* to the curve at that point.

293. In a like manner two curves may be tangent to each other (Fig. 265) when, lying in the same plane, their points of intersection become consecutive, or when a line tangent to one is at the same time tangent to the other at a common point. In such a case the normal *ac* is likewise common to both curves, the point *a* in which it intersects the tangent being termed the *point of incidence*.

294. The *normal plane* to any point of a curve is the plane led through this point at right angles to the tangent line.

295. A *point of inflexion* upon a branch of any curve is one in which the normal plane and every tangent plane at this point with one exception crosses it (Fig. 266, 1).

A *cusp point* is that in which two branches of a curve have a common tangent and are located on the same side of the normal plane (Fig. 266, 2, 3).

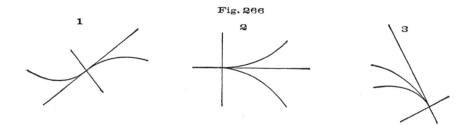

Fig. 266

1   2   3

The *angle of two curves* is the angle measured by their tangents at a common point. The curves are *tangent* when this angle is a minimum, and *normal* when it is right.

296. A rectilinear tangent to a surface is a line that touches any curve on the surface cut by a secant plane which passes through that line.

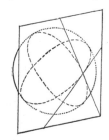

297. A plane is tangent to a surface when it has one point in common with it, or when, if any secant plane be passed through that point, the right line cut on the plane and the curve cut on the surface are tangent to each other.

Hence the tangent plane to a surface may be considered as the geometrical locus of the tangents at the point of contact.

298. A *normal* to a surface is the line drawn at right angles to the tangent plane through the point of contact, and any plane passing through a normal is a *normal plane*.

Only one normal can be drawn at any point of incidence, but an indefinite number of normal planes, all of which pass through or intersect one another in the normal line.

299. Two surfaces are tangent to each other when a plane is tangent to both at the common element of contact, or when, if any secant plane be passed through both, the line cut on the plane is tangent to the sections cut on the surfaces at a common point of contact.

300. The *angle of two surfaces* at any point is the angle of their tangent planes, or of their normals at that point. When the angle is a minimum the surfaces are tangent; when right they are normals to each other.

301. From the foregoing it follows that—

(1) Every plane passed through the point of contact on a surface cuts a section on the surface and a line on the tangent plane which are tangent to each other at that point.

(2) The number of secant planes which can be passed through the point is unlimited, giving in every case a right line on the plane and a section on the surface which are tangent to each other.

(3) All the sections thus cut have one point in common—the point of contact.

(4) All lines drawn through the point of contact tangent to the surface are lines of the tangent plane.

(5) Conversely, all lines in the tangent plane which are tangent to the surface pass through the point of contact.

(6) Tangents and normals are conjugate.

(7) Lines tangent to each other in space give projections which are tangent.

302. As the tangent plane is the locus of all tangent lines at any given point, any two tangents will determine that plane.

A normal may be determined by the tangent plane since, being perpendicular to that plane, its projections are perpendicular to the respective traces.

II.   TANGENTS AND NORMALS TO SURFACES.

303. *Ruled Surfaces.*—With single-curved surfaces the tangent plane passes through or contains a right element—*the line of contact*. If such a surface and its tangent plane be cut by any secant plane, the section on the surface and the line on the plane will be tangent to each other in a point of the line of contact.

Hence the tangent plane gives a trace tangent to the base, cut by either plane of projection, and at the foot of the line of contact.

304. A single point on a ruled surface determines a tangency, since it fixes the position of the line of contact and a second tangent line of the required plane.

305. Through a point in space a tangent plane cannot always be passed. The conditions demanded in such a case will be satisfied when, with cylindrical surfaces, the point lies beyond and towards the convex side, and, with conical surfaces when the line joining the point with the vertex passes clear of the nappes.

306. Through a line in space a tangent plane can be passed when the line is either parallel to an element or is tangent to any section cut on the surface. Under these conditions there will be one or two planes, as the line is tangent to a section or passes through the vertex of a cone, or is tangent to a section or parallel to the axis of a cylinder.

307. Parallel to a line in space two tangent planes can, in general, be passed, but parallel to a plane only one can be led to the cone and two to the cylinder.

308. In the ruled surfaces the normals have their points of incidence in the line of contact, being for any one element parallel to one another and, hence, lying in the same plane—the normal plane.

309. *Double-curved surfaces* have but one point of tangency in common with a plane.

Through a point in space an unlimited number of tangent planes may be passed to a double-curved surface, inasmuch as an infinite number of tangent lines can be drawn which, instead of forming a tangent plane, as in the cases of the ruled surfaces, determine a conic surface, through each element of which a tangent plane may be led.

310. Parallel to a line in space an unlimited number of tangent lines and planes can be passed to the surface, the lines determining a cylindrical surface and forming with their points of contact a continuous line.

Parallel to a plane in space but two tangent planes can be passed.

311. In the case of the sphere all tangent lines and planes are perpendicular to the respective radii at the points of contact; hence the traces of such a plane are always perpendicular to the projections of the radii passing through these points.

As all the radii are normals, it follows that all normal planes contain the centre and cut great circles on the surface.

With the great circles the normal surface determined by the infinite series of normal lines is a plane, but with the smaller circles it is a *cone* whose vertex is at the centre.

312. *Surfaces of Rotation.*—Through any point on such a surface a parallel and a meridian section can be cut, the rectilinear tangents to which fix the position of a tangent plane.

313. Tangents to parallels determine parallel planes, but tangents to meridians determine either a cylindrical or conical surface: cylindrical when the tangents are parallel to the axis, and conical when inclined, since, lying in the plane of the meridian, they necessarily intersect the axis in a common point.

314. The tangent plane is perpendicular to its conjugate meridian plane, and its inclination to the axis is measured by the tangent line to the meridian at the point of contact.

315. The normal surface to a meridian is a plane, but to a parallel it is a cone whose vertex lies in the axis.

316. In connection with the point, line, or plane in space the same general conditions prevail as with the double-curved surfaces.

317. PROBLEM.—*To pass a line tangent to a section of the cylinder, the point of contact being given.*

Let (b', b'') be the assumed point of contact (Fig. 240).

As the tangent line is common both to the secant and tangent planes, it is their line of intersection. Hence at the foot b'' of the element of contact draw the horizontal trace HL of the tangent plane; the point k'' in which it intersects HM is a point of the required tangent, and its vertical projection k' a point of the vertical projection $k'b'$ of that line.

318. PROBLEM.—*To pass a line tangent to a section of a cone, the point of contact being given.*

Let (h', h'') be the assumed point of contact (Fig. 241).

As in the preceding case draw the trace HL of the tangent plane, and find its intersection p'' with HM. The point so found determines the tangent $p''h''$, whose vertical projection coincides with the trace of the secant plane.

To determine the tangent in its revolved position, find the vertical piercing-point o' of the tangent line—since it remains fixed during the *rabattement* of the secant plane—and the line $o'h_1'$ is the tangent sought.

319. PROBLEM.—*To pass a line tangent to a section on a cylinder.*

Let (x', x'') be the assumed point of contact (Fig. 246).

At the foot d'' of the element of contact draw the horizontal trace HL of the tangent plane, and find the horizontal piercing-point h'' of the line of intersection between this plane and the secant plane M, the point so determined, in connection with the assumed point, fixes the position of the required tangent $(h'x', h''x'')$.

As the secant plane is brought into the horizontal plane, the piercing-point h'' remains fixed in its position; hence it will be the second point necessary to draw the revolved tangent line $h''x_1''$.

320. PROBLEM.—*To find a tangent to any plane curve not subject to geometrical definition.*

Let $A..C..F$ be any curve drawn at will, and let it be required to draw a tangent at the point D.

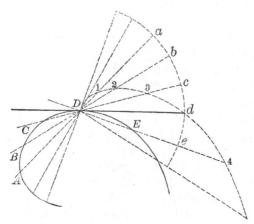

Take any number of points, as A, B, C, etc., on either side of D and draw a series of secants through them and D. Describe from D as a centre an arc of any radius, cutting the secants in points *a*, *b*, *c*, etc., and lay off on the secants distances equal to the respective chords of the original curve. Thus, make $a1 = AD$, $b2 = BD$, $c3 = CD$, $e4 = ED$, etc.; the curve traced through the points 1, 2, 3, etc., intersects the circle in the point *d* of the required tangent, since for that line the chord is reduced to a minimum.

III. PROBLEMS. TANGENTS TO RULED SURFACES.

321. PROBLEM.—*To pass a plane tangent to a right cylinder at any point of a section and determine the tangent line to that section.*

Let the cylinder be given as in Fig. 267, and let (*VM*, *HM*) be the traces of the secant plane, and (*a'*, *a''*) the given point.

The tangent plane contains the element of which (*a'*, *a''*) is a point; hence its horizontal trace *HL* is tangent to the base of the cylinder at the foot *a''*, and its vertical trace *VL* is perpendicular to *GL*.

The tangent to the section is the line of intersection between the

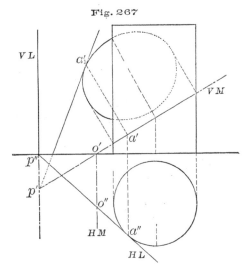

Fig. 267

tangent and secant planes (Art. 297); hence its projections coincide respectively with *VM* and *HL*. When the plane *M* is brought by *rabattement* around *VM*, the piercing-point (*p'*, *p''*) of the line of intersection or tangent remains fixed; hence the line $p'a_1'$ is the tangent required.

322. PROBLEM.—*Through a given point in space to pass a plane tangent to a cylinder.*

Let the cylinder be given as in Fig. 268, and let (*p'*, *p''*) be the given point.

As the tangent plane must pass through an element of the surface, and hence be parallel to the axis, the line (*p'o'*, *p''o''*) led through the given point will be a line of the tangent plane, and its piercing-point *o''* a point of the trace *HM*.

The vertical piercing-point n' is a point of the vertical trace VM; but as the traces intersect beyond the limits of the drawing, an additional point must be found. To effect this, pass through any point (b',

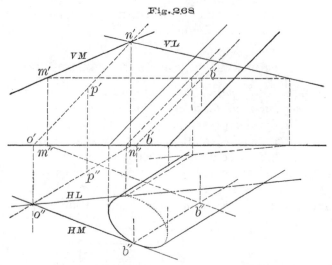

Fig. 268

b'') of the line of contact ($b'b'$, $b''b''$) a horizontal ($b'm'$, $b''m''$) of the plane M, and find its vertical piercing-point m'; the line joining n' and m' is the trace required.

A second tangent plane L may be determined by a similar construction.

323. PROBLEM.—*To pass a plane tangent to a cylinder and parallel to a given line.*

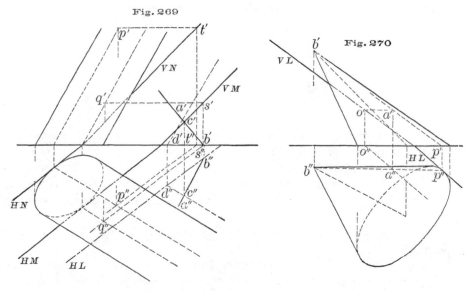

Fig. 269

Fig. 270

Let the cylinder be given as in Fig. 269, and let ($a'b'$, $a''b''$) be the given line.

From any point (c', c'') of the line let fall a parallel ($c'd'$, $c''d''$) to the axis of the cylinder, and determine the horizontal trace *HL* of the plane which contains the line so found and the given line (Art. 85).

As any plane tangent to the cylinder must be parallel to the axis, and as by the conditions of the problem it must be likewise parallel to the given line, the plane *M*, whose horizontal trace *HM* is parallel to *HL* and tangent to the base, satisfies both requirements, and hence is the horizontal trace of the required plane.

To determine the vertical trace assume any point (q', q'') of the element of contact, and draw through it a line parallel to *HM*; the line so drawn is a horizontal, and its piercing-point (s', s'') in connection with the trace *HM* fixes the position of the vertical trace *VM*.

By similar methods of construction a second plane *N* may be determined.

324. PROBLEM.—*Given one projection of a point on the surface of a cone, to pass through that point a plane tangent to the cone.*

Let the cone be given as in Fig. 270, and let a' be the vertical projection of the given point.

Pass through the point a right element ($b'p'$, $b''p''$) of the surface, and determine a'' by means of the ordinate. The tangent plane must contain this element, and hence its horizontal trace *HL* must pass through the piercing-point p'' and be tangent to the base.

To find *VL*, draw through any point of the element of contact, as (a', a''), a horizontal ($a'o'$, $a''o''$) of the plane *L*; its vertical piercing-

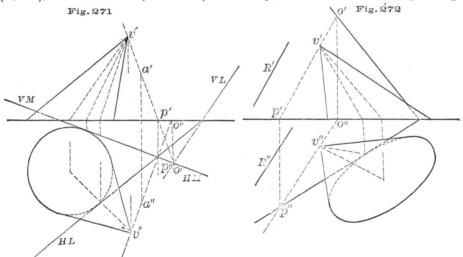

Fig. 271   Fig. 272

point o', in connection with the horizontal trace *HL*, determines the trace sought.

325. PROBLEM.—*To pass through a point in space a plane tangent to a cone.*

Let the cone be given as in Fig. 271, and let (a', a'') be the given point.

As every plane tangent to a cone passes through the vertex, the line $(a'v', a''v'')$ joining the given point and the vertex must be a line of the tangent plane, and its horizontal piercing-point p'' a point of the horizontal trace *HL* or *HM*. The vertical trace *VL* or *VM* passes through the vertical piercing-point o' of this same line.

A second point of the vertical trace may likewise be determined by means of a horizontal of the tangent plane or by the element of contact.

326. PROBLEM.—*To pass a plane tangent to a cone and parallel to a given right line.*

Let the cone be given as in Fig. 272, and let *R* be the given line.

The line $(v'p', v''p'')$ led through the vertex and parallel to the given line must be a line of the tangent plane, since every tangent plane passes through the vertex and, by the conditions of the problem, must be parallel to the given line.

Find the piercing-points of the line $(v'p', v''p'')$ and proceed as in the preceding case.

IV. SURFACES OF REVOLUTION.

327. PROBLEM.—*Through a point on a surface of revolution to pass a plane tangent.*

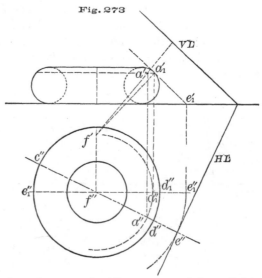

Fig. 273

Let the surface be given as in Fig. 273, and let a'' be the horizontal projection of the point.

As the tangent plane is perpendicular to the meridian plane of a surface at the point of incidence of the normal, and passes through the tangent to the meridian section at that point, through (a', a'') pass a secant plane cutting the meridian, of which $c''d''$ is the horizontal projection. Revolve this plane around the axis of the surface until it assumes a position parallel to *V*, carrying with it the point, meridian, tangent and normal; these constructive elements are respectively projected at (a_1', a_1''), $(d_1'c_1', d_1''c_1'')$, $(a_1'e_1', a_1''e_1'')$ and $(a_1'f', a_1''f'')$. By counter-rotation the foot of the tangent line falls at e'', and the normal coincides with the horizontal trace of the secant plane. As the tangent plane is perpendicular to the normal af, HL drawn through e'', and perpendicular to that line, is the horizontal trace sought.

The vertical trace VL is found by leading through the ground-point a perpendicular to the vertical projection $f'a'$ of the normal.

328. PROBLEM.—*Through a given point in space to pass a plane tangent to a sphere at a given meridian.*

Let the sphere be given as in Fig. 274, and let (a', a'') be the point, and HL the trace of the meridian plane.

Any plane tangent to a surface of revolution is tangent to every section cut on that surface by secant planes which pass through the point of contact. In the present case one of the sections is the meridian, and

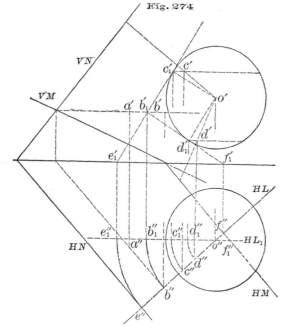

Fig. 274

a second may be assumed to be a parallel or circle passing through that same point. The rectilinear tangents to these curves are at *right angles* to each other and fix the position of the tangent plane.

Hence, through the given point (a', a'') let fall a perpendicular $(a'b', a''b'')$ upon the meridian plane, L, and find its piercing-point (b', b'') thereon. Through this latter point lead tangents to the meridian; these tangents, together with the line $(a'b', a''b'')$ which is parallel to the tangent of the circle, serve to fix the position of the tangent planes required.

To determine the tangents revolve the meridian plane around the vertical axis of the sphere until it assumes a position parallel to V, when b'' falls at b_1'' and the vertical projection of the meridian coin-

cides with the apparent contour of the surface. As the line ab is horizontal, its vertical projection $a'b'$ is parallel to GL and contains the point b_1'.
Hence $b_1'c_1'$ and $b_1'd_1'$ are the revolved tangents sought, and (c_1', c_1'') (d_1', d_1'')
their points of contact. By counter-rotation (c_1', c_1'') falls at (c', c''), (d_1', d_1'')
at (d', d''), and the feet of the tangents fall at (e'', e') and (f'', f').

As the tangent plane is perpendicular to the normal at the point of
contact, the lines HN and HM drawn through e'' and f'' respectively,
and perpendicular to the horizontal projections $o''c''$ and $o''d''$ of the
normals, are the horizontal traces sought. The vertical traces VN and
HN are found by leading through the ground-points perpendiculars to
the respective vertical projections $o'c'$ and $o'd'$ of the normals.

329. Problem—*To pass a plane tangent to a surface of revolution and
parallel to a given plane.*

Let the surface be given as in Fig. 275, and let L be the given plane.

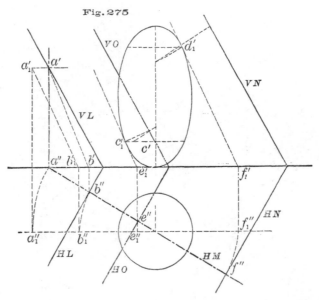

Fig. 275

Any plane tangent to the surface is perpendicular to the plane of
the meridian passing through the point of contact; hence the given
plane L and the required plane being parallel are both perpendicular
to a common meridian plane, which, according to Geometry, intersects
them in parallel lines. If, then, the plane M be passed through the
axis of the surface perpendicularly to the plane L, it will cut the required meridian on that surface and a line on the plane, to which line
the tangent to the meridian will be parallel.

To determine this tangent find the line of intersection between the
planes L and M; $(a'b', a''b'')$ are the projections required. Now revolve
the plane M around the axis of the surface until it assumes a position
parallel to V, carrying with it the line $(a'b', a''b'')$ and the meridian
section; the former falls at $(a_1'b_1', a_1''b_1'')$, and the latter coincides in
vertical projection with the apparent contour of the surface.

It is evident that the lines $c_1'e_1'$ and $d_1'f_1'$, drawn parallel to $a_1'b_1'$, are the vertical projections of the tangents in the revolved position. By counter-rotation they are projected in the horizontal trace, *HM*, of the meridian plane, their feet falling in the points f'' and e'' respectively. The horizontal traces of the required planes pass through these points and are parallel to *HL* of the given plane; the vertical traces pass through the ground points and are parallel to *VL*.

A verification of the solution will be found in the fact that the traces are perpendicular to the respective normals.

330. PROBLEM.—*To pass a plane parallel to a given line and tangent to a surface of revolution at a given meridian.*

Let the surface be given as in Fig. 276, and let $(a'b', a''b'')$ be the given line, and *HM* the horizontal trace of the meridian plane.

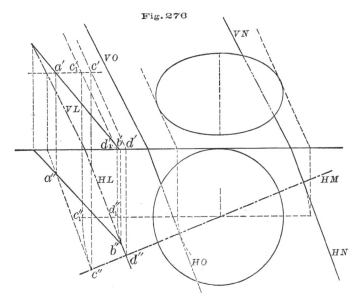

Fig. 276

The tangent plane contains the two tangents, one to the meridian section and one to the parallel or circle at the point of contact.

Hence, from any point (a', a'') of the given line draw a parallel to the tangent to the circle—that is, a perpendicular to the meridian plane; the plane *L* which passes through these two lines is parallel to the tangent plane, and cuts the plane of the meridian in a line parallel to the tangent line. The problem is, then, resolved into the preceding case. Thus, the plane *L* intersects the meridian plane *M* in the line $(c'd', c''d'')$, to which the tangent to the meridian must be drawn parallel. To effect this construction, revolve the meridian plane around the axis of the surface until it assumes a position parallel to *V*, and determine the tangent lines as in Fig. 275. By counter-rotation, the traces of the tangent planes *N* and *O* may readily be found parallel to those of the plane *L*.

V. WARPED SURFACES OF ROTATION.

331. PROBLEM.—*The axis and generatrix of a warped surface of rotation being given, to pass a plane tangent at a given point of the surface.*

Let the surface be the hyperboloid of one nappe (Fig. 277), of which $(a'b', a''b'')$ are the projections of the axis, $(c'd', c''d'')$ those of a generatrix, and e'' the horizontal projection of the given point.

As the generatrix has been assumed to be parallel to V, the perpendicular $a''f''$ let fall upon it from the centre a'' measures the shortest distance between the axis and the generatrix, or, in other words, the radius of the circle of the gorge $f''g''h''$ in its horizontal projection. The vertical

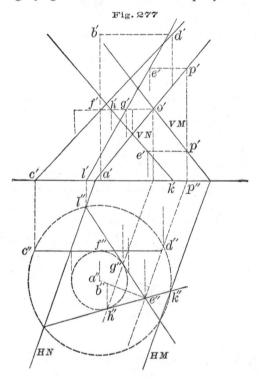

Fig. 277

projection of the circle passes through f', its assumed height above the horizontal plane. The base of the surface, or the section cut by H, is determined by the horizontal piercing-point c'' of the generatrix, whose distance from a'' measures the radius of the circle $c''l''k''$ (Fig. 134).

All the generatrices must be projected on H tangent to the circle of the gorge; hence $e''l''$ and $e''k''$ are the horizontal projections of those which pass through the given point e. The vertical projections of these generatrices pass through l' and k' and the points of tangency g' and h' in the circle of the gorge, while an ordinate from e'' determines the points e' and e', the vertical projections of the points to either of which a tangent plane is to be drawn.

As the tangent plane must contain a generatrix and also a tangent line to the parallel cut through the point of contact (e', e''), and as the line $e''p''$ drawn perpendicular to $a''e''$ is the horizontal projection of the tangents to the two parallels, their piercing-points p' and p', in connection with the piercing-points l'' and k'' of the generatrices, are sufficient to fix the traces of either tangent plane. Draw, then, the horizontal traces HM and HN parallel to $e''p''$, and let the vertical traces pass through the vertical piercing-points p' and p' respectively.

A verification of the solution will be found in the fact that, the line ($h'g'$, $h''g''$) being common to the two tangent planes, its piercing-point o' is also common to the two vertical traces.

332. PROBLEM.—*Through a given point in space to pass a plane tangent to a warped surface of rotation at a given meridian.*

Let the surface be given as in the preceding case (Fig. 278), and let (a', a'') be the point, and HL the horizontal trace of the meridian plane.

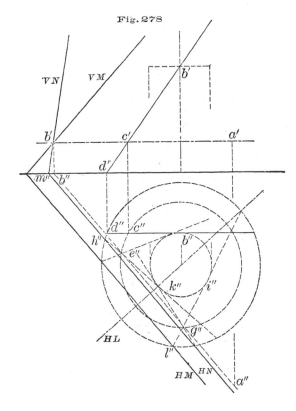

Fig. 278

As the tangent plane is perpendicular to the meridian plane, its horizontal trace is perpendicular to HL. If now an auxiliary horizontal plane be passed through the given point (a', a''), it will cut on the tangent plane a horizontal ($a''b''$, $a'b'$) whose piercing-point (b', b'') is a point of the vertical trace.

This auxiliary plane also cuts the surface in a parallel whose horizontal projection $c''e''g''$ may be found by means of the point of inter-section (c', c'') on the given generatrix $(f'd', f''d'')$, while the parallel itself is intersected by the horizontal $(a''b'', a'b')$ in two points (g'', g') and (e'', e'). But as the tangent plane must contain a generatrix having a point in common with the parallel already cut, $g''i''$ and $e''k''$ are the projections of two generatrices, either of which lies in a tangent plane that satisfies the prescribed conditions.

As all generatrices of the surface pierce H in the circumference of the base, the points h'' and l'' are points of the horizontal traces of the tangent planes. Hence, by leading through these points lines parallel to $g''e''$ the horizontal traces of the tangent planes are determined, the vertical traces of which pass through b'.

VI. TANGENT SURFACES AND ENVELOPMENT.

333. The tangency between surfaces may be divided into two general classes :

(1) Convex tangency, in which the convex surfaces touch, whether in point or line.

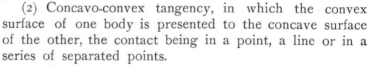

(2) Concavo-convex tangency, in which the convex surface of one body is presented to the concave surface of the other, the contact being in a point, a line or in a series of separated points.

In this latter case, if the tangency is indicated by a closed line—as, for instance, where one of the surfaces is entirely circumscribed by the other—the contact is termed an *envelopment*.

334. An enveloping surface may be generated when two lines, tangent to each other, are moved in such a way that the point of contact follows a common directrix.

If both the directrix and one of the tangent lines be right lines, then one of the surfaces generated will be a plane and the other a cylinder; if both tangents be curved lines, the surfaces generated will be cylindrical. In each case, however, the contact will be in a line parallel to the directrix.

If the directrix be a curved line, the surfaces, whatever the character of the tangent lines, will be curved, the line of contact being parallel to the directrix.

If the tangent lines be made to revolve around an axis, surfaces will be generated whose line of contact will be the circle described by the point of contact.

335. If a line tangent to a surface be so moved that its point of contact follows any plane curve of the surface, an enveloping surface will be generated.

If the generatrix be a right line moving parallel to its original position, the enveloping surface will be a cylinder; but if one point of the line be fixed, the enveloping surface will be a cone, the nature of the line of contact depending upon the character of the surface enveloped.

336. If two surfaces be made tangent to each other, and a portion of each on opposite sides of the line of contact be removed, the remaining sections will form a single surface. The operation by which this union is effected may be termed a *tangent juncture*, the line of contact now becoming the *line of juncture*.

337. The enveloping surface may be a polyhedron and the surface double-curved, in which case the contact will be in separate points, each face of the polyhedron being a tangent plane to the surface.

338. PROBLEM.—*To circumscribe a sphere by a cylinder whose axis is parallel to a given line, and determine the line of contact and the horizontal trace of the cylinder.*

Let the sphere be given as in Fig. 279, and let $(a'b', a''b'')$ be the given line.

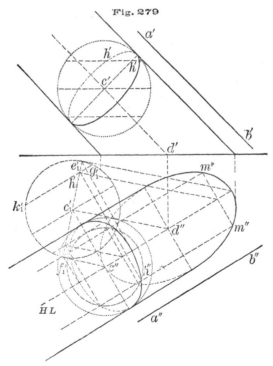

Fig. 279

Cut the sphere by a meridian plane parallel to the given line, and to the great circle thus found lead a tangent parallel to the given line. If, now, this tangent be revolved around an axis which passes through the centre of the sphere and is parallel to the given line, the point of contact will describe a great circle—the line of contact—on the surface of the sphere, and the tangent will generate an enveloping cylinder whose generatrices will be in the required position.

Thus, let the sphere be cut by the plane whose horizontal trace HL is parallel to $a''b''$, and lead through the centre of the sphere a line ($c'd'$, $c''d''$) parallel to the given line. This is a line of the meridian plane, its horizontal piercing-point d'' being in the trace. Revolve this plane around HL into H, carrying with it the great circle cut on the sphere and the axis ($c'd'$, $c''d''$) of the enveloping cylinder; the point c_1' is the centre of the sphere, and $d''c_1'$ the axis of the cylinder in the new position. Describing, then, the sphere and drawing the generatrices parallel to $c_1'd''$, $e_1'f_1'$ is the new projection of the line of contact.

To determine the primitive projections of the line of contact, the following method may be employed. Cut the surface by a series of auxiliary planes which are parallel to H. The line $k_1'g_1'$ is a section cut by such a plane, the horizontal projection of which is a circle with a diameter equal to that line. The horizontal projection h'' of the point h_1', which is common to the section and the line of contact, is readily found by means of the ordinate, while the vertical projection h' may be determined either by measuring above GL the distance of h_1' from the trace HL, or by means of the auxiliary circle passing through h_1'. A similar construction for other points of the projection $e_1'f_1'$ determines the entire projection of the line of contact.

The horizontal trace or base of the enveloping cylinder may be found by means of the horizontal piercing-points of the various generatrices which pass through the points of the line of contact above determined. Thus, the generatrices which contain the points (h_1', h'') are parallel to the axis of the enveloping cylinder, and pierce H in the points m'' and m'' of the base sought.

339. PROBLEM.—*Through a given point in space to pass a plane tangent to a surface of revolution, the point of contact being on a given parallel.*

Let the surface be given as in Fig. 280, and let ($b'c'$, $b''c''$) be the parallel, and (a', a'') the point.

Imagine a tangent to be drawn to any meridian at a point of the given parallel, and to be revolved around the surface with its point of contact constantly in that parallel; it will generate an enveloping cone whose vertex is in the axis. Any plane tangent to this cone will be tangent to the surface enveloped at a point of the line of contact or parallel. Hence the problem is resolved into Art. 325.

Should the vertex fall outside the limits of the drawing, the following method may be employed: Draw the tangent ($b'd'$, $b''d''$) to the principal meridian; this is a generatrix of the enveloping cone, the base of which is the distance of d'' from the foot of the axis. Through the given point pass a secant .plane cutting a circle on the auxiliary cone, and draw the tangents $a''m''$ and $a''n''$ to that circle; these lines are horizontals of the required tangent planes, and their tangent points m''

and n'' points in the generatrices of the cone. Hence $o''g''$ and $o''h''$ are the horizontal projections of the lines of contact, their intersections (e'', e') and (f'', f') with the given parallel being the points of contact of the tangent planes with the given surface.

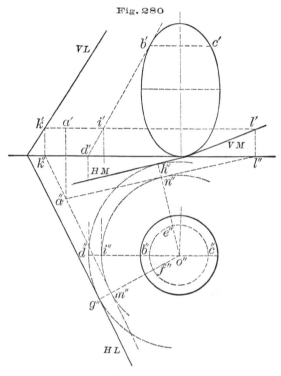

Fig. 280

The points h'' and g'', the feet of the generatrices, are the points in which the traces, HL and HM, are tangent to the base of the enveloping cone; and as the lines am and an are lines of the tangent planes, their piercing-points (k', k'') and (l', l'') determine the vertical traces VL and VM respectively.

340. PROBLEM.—*To determine a similar tangency by means of an auxiliary sphere.*

Let the surface be given as in Fig. 281, and let $(b'e', b''e'')$ be the parallel, and (a', a'') the point.

Draw a tangent $(b'd', b''d'')$ to the principal meridian, and a normal $(b'c', b''c'')$ at the point of contact; $b'c'$ is the true length of the radius of the sphere, which is enveloped by the given surface of revolution, the line of juncture being the given parallel.

Through the given point imagine a secant plane to be led, cutting a meridian section on the sphere, and further imagine two tangents to be drawn to the great circle so cut. If, now, the tangents be revolved around the line $(a'c', a''c'')$ as an axis, they will describe an enveloping cone whose line of juncture is the locus of the point of contact. Any plane which passes through (a', a'') and is tangent to the cone will be tangent to the sphere; but to be tangent to the given surface at the same time it must contain the point of intersection between the line of juncture and the given parallel.

To effect this construction rotate the plane L, carrying with it the tangents and the meridian section, around the axis of the given surface, until it assumes a position parallel to V; the trace HL then coincides with $c''b''$ and the great circle with the apparent contour $b'f_1'e'$ of the sphere, while the given point falls at (a_1'', a_1'), and the vertical projections of the tangents at $a_1'g_1'$ and $a_1'f_1'$. The line $g_1'f_1'$ is the vertical projection of the revolved line of juncture or circle of contact between

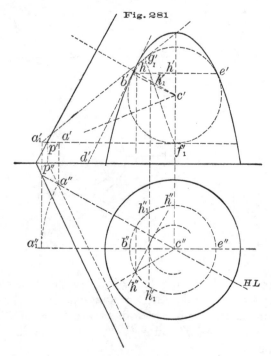

Fig. 281

the auxiliary cone and sphere, and its intersection h_1' with the given parallel is the revolved point of contact between the required tangent plane and the given surface.

Determine h_1'' and h_1''; then by counter-rotation of the plane L the line $h_1''h_1''$, which retains its perpendicular position to that plane, is projected in $h''h''$, and marks by its extremities the horizontal projections of the points of contact between the tangent planes and the given surface.

As the tangent plane is perpendicular to the normal $(c'h', c''h'')$ at the point of incidence, the traces may be determined by passing through the given point a horizontal $(a'p', a''p'')$ of that plane; its piercing-point (p', p''), in connection with the normal, is sufficient to fix the position of the tangent plane required.

CHAPTER XIV.

DEVELOPMENT.

341. If a surface be rolled upon a plane so as to bring each consecutive face in polyhedrons, or each consecutive right line in ruled surfaces in contact with that plane without stretching, folding or tearing the face or element, the surface is said to be *developed*, and the plane figure which results therefrom is termed its *development*.

342. *Rectilinear Surfaces.*—All polyhedrons are developable, inasmuch as each face, being a plane figure, may be brought in regular succession in contact with a plane, the adjacent faces, two and two, having an edge in common. As all developed surfaces necessarily represent the various parts of an object in their *true dimensions*, these must be determined from the projections either by direct calculation or by the graphical methods already explained.

343. Prismatic surfaces in which the edges are parallel present developments in which the edges are parallel.

If the surface be limited by bases at *right angles* to the axis or edges, the perimeter of the bases will be developed into *right lines* at right angles to the axis, the length of the same being equal to the sum of the sides of the bases.

With oblique prisms the bases incline to the axes, hence have sides which, as a general rule, incline to the edges and will always develop into *broken lines*.

344. Pyramidal surfaces in which the edges converge to a vertex present developments in which the edges converge, forming, two and two, the boundary lines of the plane faces.

Should the surface be limited by a base, each face will be developed as a triangle, the distance of the angular points of the base from the vertex being measured by the lengths of the respective edges of the surface and the base unrolling in a *broken line*.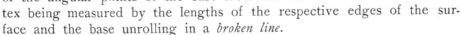

With the right regular pyramid, the edges being equal, the base will be contained within the circumference of a circle whose radius is equal to the length of an edge.

345. All points, lines and sections of a polyhedral surface preserve, after development, their relative positions on the separate faces.

346. *Ruled Surfaces.*—All surfaces in which the consecutive right elements are lines in the same plane are capable of development. All others are undevelopable.

If a cylindrical surface be rolled on a plane until it returns to the element from which it started, the surface of the plane which its successive elements have covered will be the measure of the exterior surface and, hence, its development.

If the surface be limited by bases at right angles to the axis, they will during the unrolling maintain that position to the plane of development, and hence will be developed in the traces of their planes, or as right lines perpendicular to the elements.

Should the bases be other sections of the surface, their developments will be curved lines, each point of which will be determined by the extremities of the elements cut by the secant plane.

347. In bringing the successive right elements of a conic surface in contact with a plane, the vertex remains fixed, and the extremities of the elements, should the surface be limited by a base, will mark the successive points of that base.

With the right cone of revolution, the elements being equal, the points of the base are at the same distance from the vertex, and hence, when developed, will lie in the arc of a circle whose radius is equal to the length of an element.

348. Tangents and normals to curves will be tangents and normals to their developments at the points of contact and incidence respectively.

349. PROBLEM.—*To develop a right prism with oblique section.*

Let the surface be given as in Fig. 282.

Fig. 282

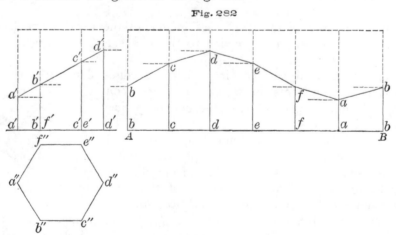

Since the base of the prism will be developed as a right line, lay off upon an indefinite line, *AB*, the sides of that base, given in their true size in *a″b″*, *b″c″*, *c″d″*, etc., and erect at the points *a*, *b*, *c*, etc., thus found a series of perpendiculars. These represent the edges of the surface, and are equal to the length as determined in the vertical projection.

Upon these lines measure the distances *aa, bb, cc,* etc., equal to those portions of the edges which are below the secant plane, and are indicated on the vertical plane in their true lengths *a'a', b'b', c'c',* etc. The broken line thus determined is the development of the section.

350. PROBLEM.—*To develop the section of a right cylinder.*

By considering the surface as a prism, whose base is a polygon of an infinite number of sides, it is evident that the principles involved in its development are precisely those employed in the preceding case.

Let the surface be given as in Fig. 283.

Divide the base into any number of small arcs, the length of the chord being approximately equal to the arc which it subtends. Upon an indefinite line lay off these chords, measuring the perimeter of the base in the line *hh.* Erect at the points *a, b, c,* etc., the perpendiculars representing the elements of the surface which pass through the extremities of the chords, and lay off upon them the distances *aa, bb, cc,* etc.,

Fig. 283

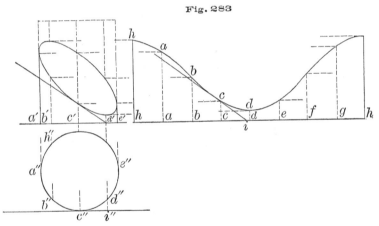

equal to the true lengths of those elements as projected on the vertical plane. The curve passing through extremities of the perpendiculars thus found is the development of the section.

To draw a tangent to the developed curve at any point c.—Lay off the distance of the subtangent *ci* equal to *c''i''*; join *c* of the developed curve and *i* for the required tangent, since the plane of the drawing is tangent at the right element passing through *c,* and hence contains both tangent and subtangent.

351. PROBLEM.—*To develop the section of an oblique cylinder.*

Let the cylinder be given as in Fig. 284.

Since the section perpendicular to the axis of a prism or a cylinder will always develop into the right line, find, by Fig. 246, the projections and true size of such a section the perimeter of which measures the width of the developed surface. Then upon an indefinite line, *AB,* lay off the distances ab, bc, cd, etc., equal to $a_1''b_1''$, $b_1''c_1''$, $c_1''d_1''$, etc., the chords of the ellipse which are approximately equal to the arcs which they subtend.

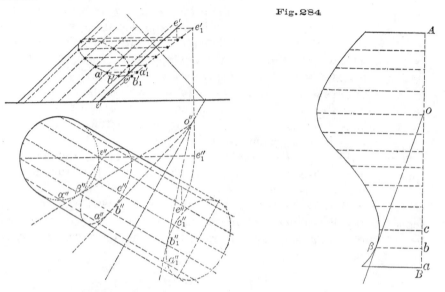

Fig. 284

To determine the distances of the points of the base from the cross-section, rotate any element as $(\epsilon''e''$, $\epsilon'e')$ parallel to V; $\epsilon'e_1'$ is its vertical projection, and measures with GL the inclination of all the elements to H. During rotation the point on any one element does not alter its distance from H; hence horizontal lines drawn from the vertical projections of the different points of the section, a', b', c', etc., determine the required distances. Thus the length of the element below the point (a', a'') is $a_1'\epsilon'$, below (b', b'') is $b_1'\epsilon'$, etc. Laying off these distances on perpendiculars to the developed section drawn through a, b, c, etc., the curve passing through their extremities is the development of the base.

To draw a tangent to the developed curve of the base at any point β. —The line will be the developed position of $o''\beta''$ tangent to the base, of which $o''b''$ tangent to the section is the subtangent; hence, making bo in the development equal to $b_1''o''$, the tangent to the section, the line joining o and β is the required tangent to the developed curve of the base.

352. PROBLEM.—*To develop a right regular pyramid and section.*
Let the surface be given as in Fig. 285.

The edges being equal, the angular points of the base lie in an arc of a circle whose radius is equal to the length of an edge.

Determine this length by rotating an edge, as $(v'd', v''d'')$, parallel to V. With $v'd_1'$ as a radius describe an indefinite arc, the centre of which is the vertex in the developed surface.

Assume an edge va, and from a lay off the chords ab, bc, cd, etc., respectively equal to the sides of the base $a''b''$, $b''c''$, $c''d''$, etc.; from the points *a*, *b*, *c*, etc., thus found draw the edges to *v*, and the figure determined is the development sought.

To develop the section, draw through the breaking-points thereof hori-

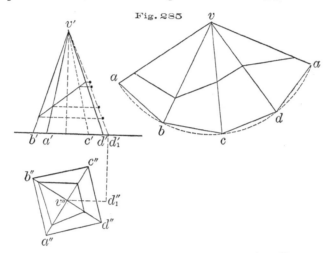

Fig. 285

zontal lines intersecting the revolved edge $v'd_1'$; the distances from v' to the points thus determined, when measured from *v* upon the developed edges, mark the points through which the developed section passes.

353. PROBLEM.—*To develop a right cone of revolution and section.*

Let the surface be given as in Fig. 286.

Fig. 286

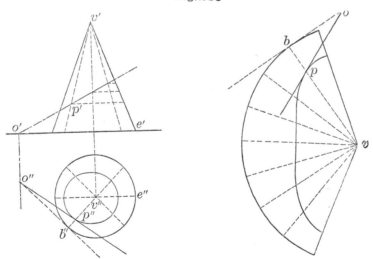

Regarding the cone as a pyramid whose base is a polygon of an infinite number of sides, the principles involved are identical with those of the preceding case.

Divide the circumference of the base into small arcs whose chords may be taken as equal to those arcs; describe by means of an element an indefinite arc, and lay off upon it the successive chords of the base.

The developed surface is the sector of a circle, in which the centre is the vertex, the radii the terminal elements of the developed surface, and the arc of the circle the measure of the perimeter of the base.

To develop the section, lead through the points in which the secant plane cuts the elements passing through the dividing points of the base, horizontal lines which intersect the element $(v'e', v''e'')$; the distances from v' to the points thus determined, when measured from v upon the developed elements, mark the points through which the developed section passes.

The tangent to any point, as p, of the developed section is found by making bo equal to the subtangent $b''o''$.

354. Problem.—*To develop the oblique pyramid and section.*

Let the surface be given as in Fig. 287.

Fig. 287

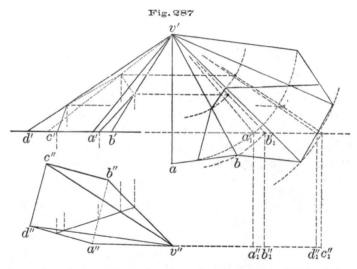

The triangular faces when developed must be exhibited in their full size; hence, determine the true lengths of the edges by rotation. The measurements thus found, in connection with the base lines, already horizontally projected in their true lengths, determine the development.

Thus, let it be required to find the face $(v'a'b', v''a''b'')$.

Rotate the edges $(v'a', v''a'')$ and $(v'b', v''b'')$ around the vertical line passing through the vertex until they assume positions parallel to V; $v''a''$ and $v''b''$ fall respectively in $v''a_1''$ and $v''b_1''$, and $v'a_1'$ and $v'b_1'$ are the vertical projections in the new position. With $v'a_1'$ and $v'b_1'$ as radii, describe indefinite arcs, and draw at will the developed edge $v'a$. From a as a centre, and with a radius equal to the side of the base of the pyramid, describe an arc; the point of intersection b between this arc and that described through the extremity b_1' completes the face vab required.

In a similar manner the remaining faces may be readily determined.

As the breaking-points of the section do not alter their heights above H during the rotation of the edges, horizontal lines drawn from their vertical projections mark at their points of intersection with the revolved edges their true distances from the vertex.

355. Problem.—*To develop the oblique cone.*

Let the surface be given as in Fig. 288.

By regarding the cone as an oblique pyramid, having for its base a polygon of an infinite number of sides, the solution will be identical with the preceding case.

An elegant solution may likewise be effected by means of the intersecting sphere, as follows:

From the vertex of the cone as a centre describe any sphere intersecting the surface of the cone in a line all of whose points are necessarily

Fig. 288

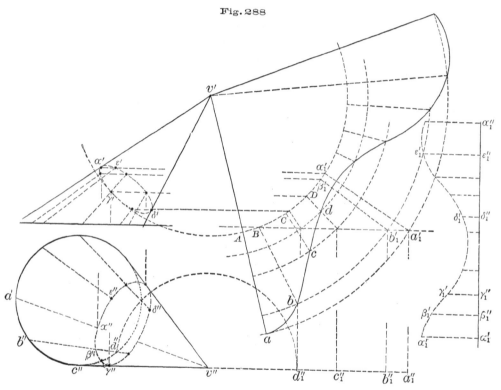

at the same distance from the vertex. Determine this line of intersection by passing through the cone a series of vertical secant planes which cut right elements on that surface and great circles on the sphere, and rotate the planes and sections parallel to V around a vertical axis passing through (v', v''). Thus, the element $(v''a'', v'a')$ falls at $(v''a_1'', v'a_1')$, the section of the sphere, in each case, coinciding with the primitive vertical projection. Drawing, then, a horizontal line through a_1', its intersection with the element passing through (a', a'') of the base determines a point a' of the required line. By a similar procedure the remaining points necessary to complete the entire line of intersection may be found.

As the line of intersection is not a plane curve, it cannot be rotated parallel to either plane of projection; the true length of the perimeter may, however, be obtained by regarding the horizontal projection $\alpha''\beta''\gamma''\delta''$ as the base of a right cylinder.

Develop this imaginary cylinder; its base is the right line α_1'' . . . δ_1'' . . . α_1'', from which the development of the line of intersection is effected by laying off the heights of its respective points above H.

Again, as the line of intersection has each point at the same distance from the vertex, describe an indefinite arc with a radius equal to that of the auxiliary sphere, and measure upon this arc the distances AB, BC, . . . equal to $\alpha_1'\beta_1'$, $\beta_1'\gamma_1'$ . . . of the developed line of intersection. Through the points A, B, C, etc., thus determined draw the elements of the cone, and lay off below the developed line of intersection the additional lengths of the elements necessary to complete the conic surface. Thus, from A lay off the distance Aa equal to $\alpha_1'\,a_1'$, found in its true size on the element in its revolved position, from B the distance $\beta_1'b_1'$, and so on for the remaining points of the base.

356. PROBLEM.—*To develop the icosahedron.*

Let the surface be given as in Fig. 289.

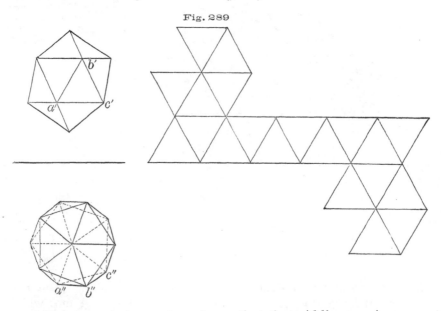

Fig. 289

An inspection of the surface shows that the middle zone is composed of a series of ten equal triangles, the length of the sides being equal to the edge $a'b'$ projected on V in its true size, and the base-lines forming two regular pentagons parallel to H. By development these two polygons become two parallel lines, between which lie the ten equilateral triangles.

As the polygons are likewise the bases of two pyramids whose faces are equilateral triangles, develop the pyramids, keeping one side of each base in common with the development of the zone.

CPSIA information can be obtained
at www.ICGtesting.com
Printed in the USA
BVHW061444150722
642051BV00002B/75